Understanding Quantum Science

Students are naturally drawn to quantum science by the intriguing behaviors of small particles. However, they can also be intimidated by the lengthy and complicated treatment found in the classroom. ***Understanding Quantum Science: A Concise Primer for Students of Chemistry, Biochemistry, and Physics*** is a highly accessible book that offers students an opportunity to grasp the most fascinating of quantum topics, without the intimidation. To be sure, math is necessary, but it is introduced as needed and kept concise. The emphasis is on the science: a certain differential equation can be solved, and when it is, we find the energies that hydrogen atom electrons are allowed to have. Each concept is developed in this manner, keeping focus on how and why it arises, and on the intriguing consequences.

This book provides a brief tour of some of the wonders of quantum science. But more than that, it is designed to be the most concise tour possible that truly explains how these wonders arise so that you can develop a working understanding of quantum concepts. If your goal is loftier and you wish to become a quantum specialist, the conceptual groundwork presented here, along with rationalization of the mathematics required, will position you well for higher level classes.

Understanding Quantum Science

A Concise Primer for Students of Chemistry, Biochemistry, and Physics

Steven M. Pascal

CRC Press is an imprint of the
Taylor & Francis Group, an **informa** business

Original illustrations by Mark Pascal and Steven Pascal

First edition published 2023
by CRC Press
6000 Broken Sound Parkway NW, Suite 300, Boca Raton, FL 33487-2742

and by CRC Press
4 Park Square, Milton Park, Abingdon, Oxon, OX14 4RN

CRC Press is an imprint of Taylor & Francis Group, LLC

Library of Congress Cataloging-in-Publication Data
Names: Pascal, Steven M., author.
Title: Understanding quantum science : a concise primer for students of
chemistry, biochemistry and physics / Steven M. Pascal.
Description: First edition. | Boca Raton, FL : CRC Press, 2023. |
Includes bibliographical references and index.
Identifiers: LCCN 2022052992 (print) | LCCN 2022052993 (ebook) |
ISBN 9781032410838 (pbk) | ISBN 9781032410852 (hbk) | ISBN 9781003356172 (ebk)
Subjects: LCSH: Quantum theory.
Classification: LCC QC174.12 .P377 2023 (print) |
LCC QC174.12 (ebook) | DDC 530.12–dc23/eng20230302
LC record available at https://lccn.loc.gov/2022052992
LC ebook record available at https://lccn.loc.gov/2022052993

ISBN: 9781032410852 (hbk)
ISBN: 9781032410838 (pbk)
ISBN: 9781003356172 (ebk)

DOI: 10.1201/9781003356172

Typeset in Times
by Newgen Publishing UK

Access the Support Material: https://www.routledge.com/9781032410852

To my P Chem students at Old Dominion University,
who ask me so many insightful questions.

Contents

The Basics

One-Dimensional Potentials

Approximation Methods

Three-Dimensional Space: Atoms and Molecules

Preface

This book arose from a six-week course of lectures that I give to my Physical Chemistry students at the start of the fall semester. Our topic is quantum chemistry, and I would like them to understand quantum science reasonably well before we apply the concepts to atoms and molecules. Sure, you could start right off with atomic orbitals, but *why* do electrons occupy orbitals, and *why* are the orbitals labeled via $\boldsymbol{n}$, $\boldsymbol{\ell}$, and $\boldsymbol{m}$ quantum numbers? Full derivation of all the necessary equations is not practical. But treating it as a black box ("because Dr. Pascal said so") is not very illuminating either. There are many excellent physical chemistry texts that discuss these topics. However, the formality required for a full year physical chemistry textbook that might contain 1000 pages can make even the best text seem stifling to a curious but uninitiated student.

So, over the course of my years of first teaching biophysics to medical students, next modern physics in a Physics department, and then physical chemistry to students majoring in chemistry or biochemistry, I developed a method to present the fundamental concepts of quantum science to students of varied backgrounds, in a relatively pain-free manner. I aimed to keep the tone conversational, and yet to retain the mathematics needed to understand the principles. Students can understand the radial and angular nodes in a 3p-orbital much better once they have seen in detail how the nodes arise in a square well. Then, they already know that more nodes mean more energy. And they don't just know it, they feel it. Science involves both knowledge and sensation. You should develop not just skills, but intuition.

This intuitive approach (note: some math is required to develop intuition) can be seen in the Table of Contents. The Schrödinger equation is the key (Chapter 2), but where did it come from, what does it mean (Chapter 3)? This exploration leads naturally to a discussion of operators, uncertainty, and quantization (Chapter 4). We next need to separate the time and space variables, so that we cand spend the rest of our time focusing on the space variables using the time-independent Schrödinger equation (TISE) (Chapter 5). Seeing how the separation of x and t is performed makes separation of the three-dimensional TISE into r, θ, and ϕ equations seem obvious (Chapter 14), which then leads logically into a discussion of angular momentum, spin, electron orbitals, and the atomic quantum numbers (Chapters 15, 16). But first, we use simple one-dimensional systems such as step functions (Chapter 7), the square well (Chapter 8), and the harmonic oscillator (Chapter 10) to develop basic quantum principles, which are summarized in Chapters 6 and 9. Chapters 11–13 show how to approximate solutions for the TISE, since exact solutions are few and far between; in fact, atomic orbitals cannot be exactly derived for any atom larger than hydrogen. Approximation is absolutely vital. And if you learn it well in one dimension, then applying perturbation theory or the Variational Method to three-dimensional multi-electron atoms will seem intuitive.

No attempt is made in this brief text to show all of the consequences of these principles, not even for atoms and molecules, let alone for condensed matter, electronic devices, or elementary particles. Some of these connections are simply stated in Chapter 17. The point is that by the end of 16 brief Chapters (perhaps six weeks), the student should be ready to learn to apply these concepts to whatever system they wish. And to understand why they are doing it, and why it works.

About the Author

Steven M. Pascal is Professor of Chemistry and Biochemistry at Old Dominion University. He has previously taught medical students at the University of Rochester Medical Center and physics students at Massey University in New Zealand. His previous book entitled *NMR Primer: An HSQC Approach* presents the basics of NMR needed by biological spectroscopists in a rigorous but mostly visual manner.

The Basics

1 Introducing Quantum Mechanics

1.1 WHAT IS QUANTUM MECHANICS?

The subject of *quantum chemistry* seeks to understand the behavior of atoms and molecules. This understanding reduces to describing the interactions between groups of electrons and nuclei. Due to the small size of the particles, *quantum mechanics (QM)* must be used instead of *classical mechanics (CM)*. For our purposes, it can be assumed that the motion of the particles is slow relative to the speed of light. Therefore, *non-relativistic QM* will be used, mainly to study how electrons "move" around the nucleus in atoms or around multiple nuclei in molecules.

Of course, this is an over-simplification in several ways. First, while electrons and nuclei can at times be treated like particles, they are *quantum* particles. That means that we typically do not, and cannot, know exactly where they are and how fast they are moving. Instead, we calculate the probability that they are in some region of space at some particular time. This probability is expressed in terms of a wave function, denoted $\Psi(x,t)$.

All particles – including the earth, for instance – are subject to the laws of QM. We could in theory derive a wave function Ψ describing the orbit of the earth around the sun. However, for large particles, QM equations reduce to CM equations. CM can be thought of as a shortcut form of QM, to be applied to the large objects we normally observe. Therefore, we needn't calculate the probability that the earth will be at a particular place: its position is known with great certainty relative to its size. For atoms and molecules, however, this is not the case, and the uncertainty in the position of an electron is typically many times its own "size". Hence the need for QM to understand atoms and molecules.

The QM behavior of electrons, atoms, and molecules is not merely an inconvenience that complicates calculations. It also leads to unexpected behaviors that are critical to the structure of the universe. For instance, the *Pauli exclusion principle* states that no two electrons of an atom can have the same quantum state ($\boldsymbol{n}$, $\boldsymbol{\ell}$, $\boldsymbol{m_l}$, $\boldsymbol{m_s}$). Without this fact, all electrons, even in a large atom, could pile into the 1s orbital, and none of the rich chemistry that we know would exist. The Pauli exclusion principle is but one important consequence of the laws of QM.

DOI: 10.1201/9781003356172-2

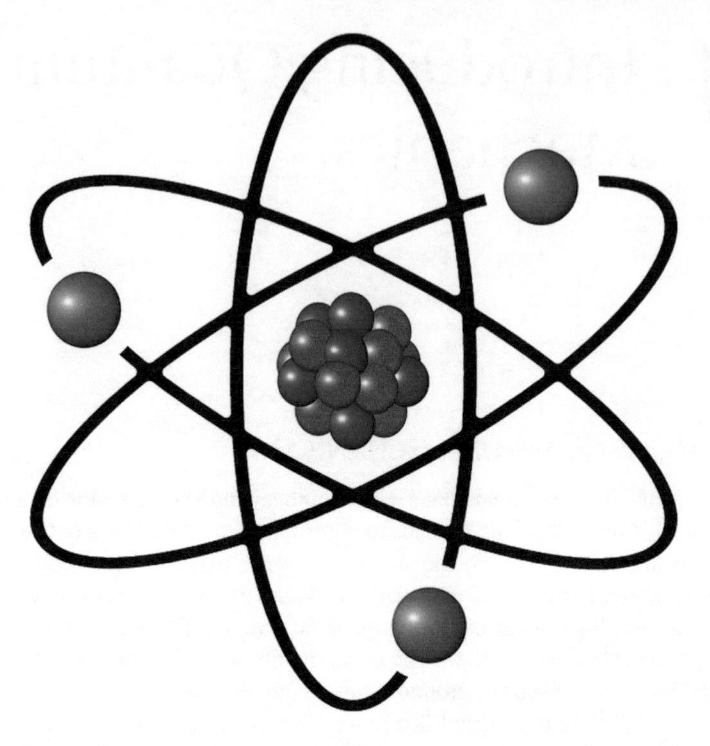

FIGURE 1.1 Innacurate depiction of a lithium atom. [Image by © Gerd Altmann, https://pixabay.com/illustrations/atom-symbol-character-abstract-68866/]

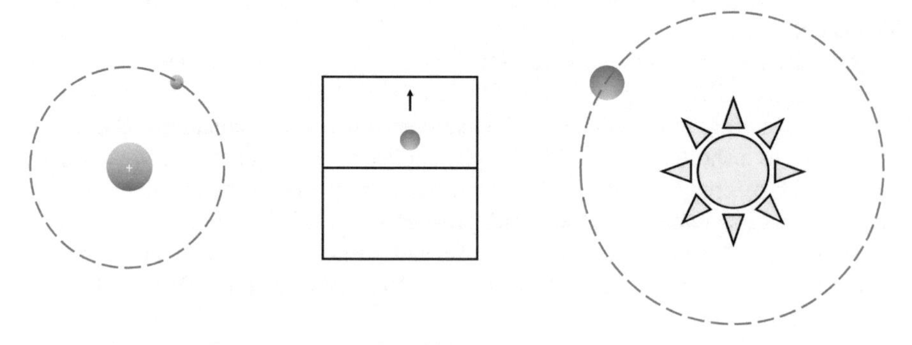

FIGURE 1.2 All confined motion is quantized. The motion of an electron around a nucleus, a squash ball on a squash court, or the earth around the sun are all quantized.

PROBLEMS

1.1 Briefly define these quantum phenomena: **(a)** Pauli exclusion principle, **(b)** Aufbau principle, **(c)** Hund's rule, **(d)** Heisenberg uncertainty principle, **(e)** Schrödinger's cat, **(f)** wave-particle duality, **(g)** photoelectric effect, **(h)** Ehrenfest's theorem, **(i)** the correspondence principle.

2 The Schrödinger Equation

In order to find Ψ, and therefore the probability of an electron to be at some position x at some time t, the *time-dependent Schrödinger equation (TDSE)* is used:

$$\text{TDSE (free space): } [-\hbar^2/2m(\partial/\partial x)^2]\ \Psi = i\hbar(\partial/\partial t)\ \Psi \qquad (2.1)$$

2.1 COMPARISON WITH THE CLASSICAL WAVE EQUATION

Where did the TDSE equation come from, and what does it mean to "solve" it? To understand, first note that the TDSE looks a bit like the *classical wave equation*, say for the vibration of a guitar string:

$$\text{Classical wave equation: } (\partial/\partial x)^2F = 1/c^2(\partial/\partial t)^2F \qquad (2.2)$$

A guitar string can be thought of as a line (line segment) lying along the x-axis. Equation 2.2 simply states that the displacement F (height above resting position) of each part of the guitar string, at any time, can be described by a function F(x,t). F is thus a function of both x and t. The equation states that the double derivative of F with respect to x equals the double derivative with respect to t, apart from some constant c^2. To "solve" this equation, we must find a function F(x,t) that makes the equation true. One possible answer is $F = \sin(kx–\omega t)$. Plugging this form of F into equation 2.2 and differentiating produces:

$$-k^2\sin(kx-\omega t) = -(1/c^2)\omega^2\sin(kx-\omega t) \qquad (2.3)$$

Cancelling like terms on either side reduces this equation to $\omega^2 = c^2k^2$. Therefore, $F = \sin(kx–\omega t)$ "solves" the classical wave equation, so long as ω and k are related by a factor of c.

DOI: 10.1201/9781003356172-3

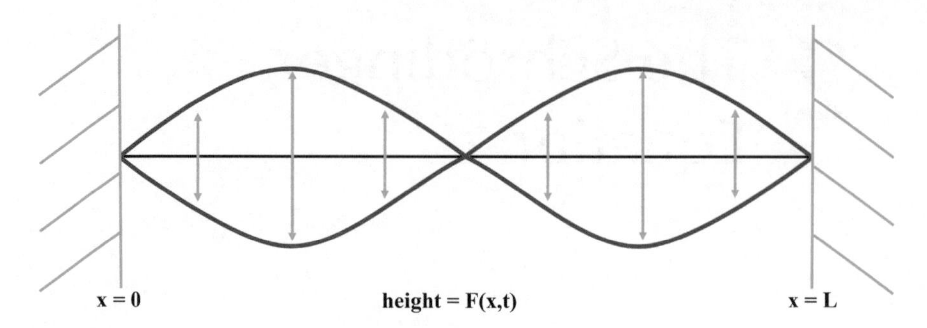

FIGURE 2.1 Standing wave on a guitar string.

2.2 COMPLEX SOLUTION TO THE CLASSICAL WAVE EQUATION

Note that $F = \cos(kx\text{-}\omega t)$ is also a solution to equation 2.2 (plug it in to confirm). In addition, $F = e^{i(kx-\omega t)}$ is a solution. In this function, $i = \sqrt{-1}$. Plugging this exponential function into equation 2.2 shows that it satisfies the wave equation, again only if $\omega^2 = c^2k^2$.

Why are there three different types of solutions possible (sin, cos, exponential)? The sin and cos solutions are basically the same: both waves look the same, but one is shifted or delayed relative to the other. We needn't here concern ourselves with this shift.

The exponential form is something quite different: it is a complex solution. Here "complex" means it has a real part (no *i*) and an imaginary part (with *i*). In order to use complex numbers, we need to know only a few facts, which will be discussed as needed. For now, just consider $e^{i\Theta}$ to be a simple shorthand notation for $\cos\Theta + i\sin\Theta$. Therefore:

$$e^{i(kx-\omega t)} = \cos(kx-\omega t) + i\sin(kx-\omega t) \tag{2.4}$$

The $\cos(kx\text{-}\omega t)$ term is the real part, while $i\sin(kx\text{-}\omega t)$ is the imaginary part. While this complex exponential solves the equation, we don't really need both parts. The motion of the guitar string is described by a sin wave or cos wave, not by their combination in a complex exponential. The only reason the exponential form might be used in the case of a guitar string is that taking the derivative of an exponential is simpler than taking the derivative of sin or cos. The true solution for the guitar string can be thought of as just the real part of the exponential.

2.3 WHAT ARE k AND ω?

The constants k and ω are multiplied respectively by x and t in the above wave functions. Since ω multiplies t, it is the *angular frequency* of the wave: how fast the height of the wave changes with time. By analogy, k, which multiplies x, is a *spatial frequency*, determining how fast the wave height changes as we move along x. Clearly, k is related to the wavelength, λ. To be precise, $k = 2\pi/\lambda$.[1] With this inverse

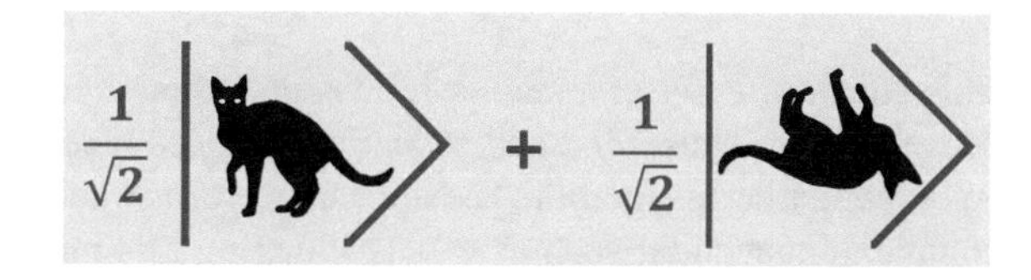

FIGURE 2.2 Erwin Schrödinger is perhaps most famous for his concept of a cat, hidden in a box, that exists in a superposition of dead and alive states. This concept is related to the Schrödinger wave equation. [LEFT figure by © Николай Антоненко, https://www.neatoshop.com/product/Schrodingers-Cat-Dead-And-Alive-Cat-Lover, by permission; RIGHT figure uses a Black cat image by © Nevit Dilmen, https://commons.wikimedia.org/wiki/File:Black_Cat_02812_svg_vector_nevit.svg, licensed under CC BY-SA 3.0. Image was duplicated and rotated, and red labeling was added.]

relationship to λ, it becomes clear that k is essentially the number of waves that will fit in a particular space. For this reason, k is called the *wave number*.

2.4 WHY DOES AN ELECTRON NEED A WAVE EQUATION?

To review, F(x,t) represents the height of the guitar string, at each position x, at each time t, while ω and k describe the wave frequency and wave length, respectively. The function F is thus a comprehensive description of the guitar string. By analogy, what does Ψ(x,t) represent for the electron? The electron is a particle, not a string. We don't need a wave function to describe the position of different parts of an electron, like for different parts of a string, do we?

Well, in a way, we do. Recall that the electron has certain probabilities to be found in various places. We can think of that probability as a wave: it is high in some places and low in others. The electron, or at least its probable location, can be thought of as being spread out over space, sort of like a guitar string. An electronic orbital is an example of this sort of spatial distribution: the electron might be found anywhere in the orbital.[2]

The function Ψ is found by "solving" the TDSE; that is, by coming up with (perhaps by guessing) some function Ψ(x,t) that can be plugged into the TDSE (equation 2.1) to satisfy the equality. In other words, to understand the behavior of an electron, we must "solve" the TDSE.

2.5 WHAT'S DIFFERENT ABOUT THE SCHRÖDINGER EQUATION?

The similarities between the TDSE and the classical wave equation are obvious, but there are also differences, most notably that (a) the TDSE has only a single time derivative; and (b) the constant c is replaced by some combination of ħ, 2, m, and *i*. The single time derivative means that Ψ cannot be a simple sin function like Ψ = sin(kx-ωt). If a sin function is plugged into equation 2.1, the result is:

$$\text{constant} \cdot \sin(kx-\omega t) = \text{constant} \cdot \cos(kx-\omega t) \qquad (2.5)$$

This equation is false: a sin wave does not equal a cos wave. Instead, Ψ here must be a function whose first and second derivatives take the same form as each other. An exponential fits the bill. The appearance of *i* in the TDSE suggests that we try a complex exponential such as $e^{i(kx-\omega t)}$. Plugging this into equation 2.1 and differentiating gives:

$$(\hbar^2k^2/2m)\, e^{i(kx-\omega t)} = (\hbar\omega)\, e^{i(kx-\omega t)} \qquad (2.6)$$

On dividing both sides by the exponential, this reduces to:

$$\hbar^2k^2/2m = \hbar\omega \qquad (2.7)$$

Therefore, the function $e^{i(kx-\omega t)}$ solves the TDSE (equation 2.1) so long as $\hbar^2k^2/2m = \hbar\omega$.

Note: the solution for the TDSE in free space is necessarily complex: it is not just one option that we are free to ignore, as it was for the guitar string. This will be true in general: in quantum mechanics, the solution is often *required* to be a complex function.

2.6 IMAGINARY PROBABILITY?

We have "solved" the TDSE in free space, but the solution $\Psi = e^{i(kx-\omega t)}$ is complex. If Ψ represents the probability to find the electron at position x at time t, how can it be complex? Can a probability be complex?

No, it cannot. The resolution of this conundrum is that the probability P(x,t) to find the particle at position x at time t is not $\Psi(x,t)$, but rather the square modulus [3] of $\Psi(x,t)$:

$$P(x,t) = \Psi^*(x,t)\Psi(x,t) = |\Psi|^2 \quad \text{[probability]} \qquad (2.8)$$

2.7 REPLACING k AND ω WITH p AND E (FOR PHOTONS)

Einstein's explanation of the *photoelectric effect* shows that photons of frequency ω have an amount of energy:

$$E = \hbar\omega \qquad (2.9)$$

Similarly, Einstein's *theory of relativity* shows that a photon has a wavelength λ related to its momentum p by:

$$\lambda = h/p \qquad (2.10)$$

Given that $k = 2\pi/\lambda$, it is easily shown that:

$$p = \hbar k \tag{2.11}$$

These equations each contain Planck's constant h. The symbol $\hbar$ stands for h divided by 2π.

2.8 EXTENDING THESE TWO EQUATIONS TO ELECTRONS

This becomes relevant to our study of electrons, if we assume that the above three relationships (eqs. 2.9, 2.10 and 2.11) apply not only to photons, but also to electrons. The assumption that $\lambda = h/p$ (and hence $p = \hbar k$) holds also for massive particles (those with mass) was made by de Broglie, hence it is called the *de Broglie hypothesis*. Equation 2.11 is merely a rearrangement of the de Broglie hypothesis. There is no common name for the assumption that $E = \hbar\omega$ for massive particles.

These assumptions require an intuitive leap: light was first known to act as a wave. Scientists later realized that light can also be thought of as containing particles (photons), where each particle has an energy and momentum defined by the frequency and wavelength of the light (eqs. 2.9 and 2.10, respectively). Conversely, electrons were first thought of as particles. Scientists later realized that electrons can also be thought of as (probability) waves. The intuitive step was to assume that equations 2.9 and 2.10, originally derived to relate photons to light waves, also pertain to electrons and their probability waves. Under this assumption, the E and p of an electron can be "read" simply by taking the values of ω and k in its wave function $\Psi(x,t)$, and multiplying by $\hbar$. Now we see why the TDSE is so valuable!

Summary

The Schrödinger equation is analogous to the classical wave equation. Its solutions $\Psi(x,t)$ describe the position of a particle at various times. It differs from the classical equation in that the solution must be squared to provide the probability to find the particle at each position. In addition, the solution may be complex, although the complexity vanishes during the squaring procedure. Through this equation, the probability distribution of a quantum particle can be determined, along with its wavelength and frequency. The distribution can also be used to calculate the particle's momentum and energy.

PROBLEMS

2.1 Assume that the A string on a six-string guitar has a vibrational frequency of $\nu = 110$ Hz, and the length of the string is 60 cm. Find the speed that a disturbance can move along the string (c) Hint: assume a fundamental vibration of the string; that is, assume that ½ wavelength fits onto the string.

2.2 Use equations 2.7 and 2.9 to solve for k in terms of E.

2.3 Show that the de Broglie hypothesis equation ($\lambda = h/p$) was already known to be true for photons by using the following equations that were established for photons: $E = pc$ (from relativity), $E = h\mathbf{v}$ (from the photoelectric effect), and $c = \lambda v$ (a standard equation relating the speed of a wave to its wavelength and frequency).

2.4 Show that $p = \hbar k$, using the de Broglie equation and the definition of wave number k ($k = 2\pi/\lambda$).

NOTES

1 The angular frequency ω is related to frequency v by $\omega = 2\pi v$. Thus $\omega = 2\pi/T$, where T is period. This latter equation is very similar to the equation $k = 2\pi/\lambda$, and helps to show the similarity between k and ω.

2 Technically, an electron even has some probability to be found outside of the "orbital" to which it is assigned. The usual drawing of an orbital simply depicts the region of space in which the electron is most likely to be found.

3 The process of multiplying a complex number or function by its own complex conjugate always produces a real quantity. Thus, the probability P(x,t) is always real. The complex math you need is that (a) the complex conjugate of Ψ, denoted by Ψ^*, is found simply by changing all i to $-i$; and (b) $i^2 = -1$. For brevity, the square modulus $\Psi^*\Psi$ is often written $|\Psi|^2$ and pronounced "psi squared".

3 Deriving the Schrödinger Equation

Clearly the TDSE is valuable for analyzing the electron, but how was it "derived"? A true derivation is elusive: it was postulated moreso than derived. However, the TDSE can be surmised via the following argument. Briefly, the steps are:

- assume that a free electron can be described by a complex wave $\Psi = e^{i(kx-\omega t)}$
- show that derivatives of Ψ with respect to x or t produce a factor of k or ω
- use QM to exchange k and ω for momentum p and energy E, respectively
- use $E = p^2/2m$ to relate one t derivate to two x derivatives, forming the TDSE
- derive the general TDSE by adding potential energy V to both sides of the equation

In a bit more detail:

- assume that a complex wave $\Psi = e^{i(kx-\omega t)}$ is appropriate to describe the probable position of a free electron
- assume that a wave equation, somewhat resembling equation 2.2, is associated with Ψ.
- to seek the form of this wave equation, examine the first derivatives of Ψ with respect to x and t:

$$(\partial/\partial t)\,\Psi = -i\omega\,\Psi \quad \text{multiply by } i \text{ to get the slightly simpler} \quad i(\partial/\partial t)\,\Psi = \omega\Psi \tag{3.1}$$

$$(\partial/\partial x)\,\Psi = ik\,\Psi \quad \text{divide by } i \text{ to get the analogous} \quad (1/i)(\partial/\partial x)\,\Psi = k\Psi \tag{3.2}$$

- multiply the above two equations by $\hbar$ and substitute $p = \hbar k$ and $E = \hbar\omega$:

$$\hbar\, i(\partial/\partial t)\,\Psi = E\Psi \tag{3.3}$$

DOI: 10.1201/9781003356172-4

$$(\hbar/i)(\partial/\partial x)\,\Psi = p\Psi \qquad (3.4)$$

- these two equations tell us that we can't use the classical wave equation (equation 2.2) for the electron, because equating double derivatives with respect to x and with respect to t would result in equating E^2 with p^2, apart from some constants
- there *is* a relationship between E and p for a free particle (V = 0). Start from $E = ½\, mv^2$. In terms of momentum, this can be rewritten as $E = p^2/2m$ (problem 3.1). This suggests that one time derivative (producing a factor of E) equals two space derivatives (producing a factor of p^2)
- our surmised wave equation now looks like the following:

$$(\partial/\partial x)^2\,\Psi = (\partial/\partial t)\,\Psi \qquad (3.5)$$

- plugging $\Psi = e^{i(kx-\omega t)}$ into this equation yields:

$$-k^2\,\Psi = -i\omega\,\Psi \text{ (dividing by } -\Psi \text{ yields } k^2 = i\omega) \qquad (3.6)$$

- this equation ($k^2 = i\omega$) is not true. It is true that $p^2/2m = E$, or, using equations 2.9 and 2.11, $\hbar^2k^2/2m = \hbar\omega$. So equation 3.5 becomes true if we multiply on the left by $-\hbar^2/2m$ and on the right by $i\hbar$:

$$[-\hbar^2/2m(\partial/\partial x)^2]\,\Psi = i\hbar\,(\partial/\partial t)\,\Psi \qquad (3.7)$$

- prove to yourself that this equation works by plugging $\Psi = e^{i(kx-\omega t)}$ into it
- equation 3.5 is the TDSE in free space (identical to equation 2.2). Apparently, it simply states that KE (kinetic energy) = energy. This is true in free space, where V = 0. We can generalize to the case of an electron subject to a potential V by noting that E = KE + potential energy (V). This fact can be incorporated into the TDSE simply by adding V to the left-hand side of equation 3.7:

$$[-\hbar^2/2m(\partial/\partial x)^2 + V]\,\Psi = i\hbar\,(\partial/\partial t)\,\Psi \qquad (3.8)$$

This is the TDSE in one dimension for a particle subject to a potential V.

The remainder of this book springs from equation 3.8. Chapter 4 introduces concepts such as operators and eigenvalues as they pertain to the TDSE. In Chapter 5, the TDSE is separated into space and time parts. General rules for solutions to this equation are discussed in Chapter 6. From that point on, we will primarily focus on the spatial part of the TDSE, which we designate the TISE, for *time-independent Schrödinger equation*. The TISE is solved for the case of a few simple potentials V in Chapters 7, 8, and 10. Chapter 9 presents the laws of QM (formal postulates). Methods to find approximate solutions to the TISE are described in Chapters 11–13. The TISE is expanded to three dimensions in Chapter 14, which gives rise to the concept of angular momentum (Chapter 15) and ultimately to electronic orbitals (Chapter 16). Finally, the use of these principles to understand multi-electron atoms, molecules, and spectroscopy is briefly introduced in Chapter 17.

FIGURE 3.1 A quote by Schrödinger.

[Photo from https://commons.wikimedia.org/wiki/File:Erwin_Schrödinger_(1933).jpg, public domain]

PROBLEMS

3.1 Starting from KE = ½ mv^2, show that KE = $p^2/2m$.

3.2 Prove that $\Psi = e^{i(kx-\omega t)}$ solves the TDSE for a free particle (equation 3.7). Also show the conditions under which it solves this equation. That is, solve for the value that k must have in order to solve the TDSE. Compare your answer to the answer for Problem 2.2. Hint: you may find it easier to use the form $\Psi = e^{ikx}\, e^{-i\omega t}$.

3.3 The potential energy V appears on the LHS of equation 3.8, but not on the RHS. Why is it missing from the RHS?

4 Operators, Oscillations, Uncertainty, and Quanta

4.1 OPERATORS

During the derivation of the Schrödinger equation in Chapter 3, we incidentally encountered two "operators". Acting on Ψ with $i\hbar\,(\partial/\partial t)$ produced a factor of E. Acting on Ψ with $\hbar/i(\partial/\partial x)$ produced a factor of $\hbar k$, which equals p. Therefore, we can define the energy ($\hat{E}$) and momentum ($\hat{p}$) operators, which "extract" the value of E and p from the wave function:

$$\hat{E} = i\hbar\,(\partial/\partial t) \quad \hat{p} = \hbar/i(\partial/\partial x) \tag{4.1}$$

Using these two operators, the following two "eigenvalue" equations can be written. The two equations below show that we can operate on Ψ (the eigenfunction) and get the same Ψ back, with a factor in front (the eigenvalue):

$$\hat{E}\Psi = E\Psi \quad \hat{p}\Psi = p\Psi \tag{4.2}$$

"Eigen" is German for "special". Only certain special functions Ψ satisfy these eigenvalue equations. If it does satisfy, then Ψ is an eigenfunction of that operator. For instance, $\Psi = e^{i(kx-\omega t)}$ satisfies both equations in equation 4.2, and so it is an eigenfunction of both operators, with eigenvalues $\hbar\omega$ and $\hbar k$, respectively (problem 4.1). On the other hand, $\Psi = \cos(kx-\omega t)$ satisfies neither equation.

Operators are vital in QM. They can be used to directly analyze the electron, as shown in the above paragraph. In fact, the TDSE is an operator equation. It can be rewritten as:

$$\mathcal{H}\Psi = i\hbar\,(\partial/\partial t)\Psi \tag{4.3}$$

where $\mathcal{H}$ is called the Hamiltonian operator, and just represents $[-\,\hbar^2/2m(\partial/\partial x)^2 + V]$. The symbol $\mathcal{H}$ will often be used in place of this longer notation, to save ink (and time). Note that the Hamiltonian $\mathcal{H}$ is simply another type of energy operator that "extracts" the kinetic energy (first term) plus the potential energy.

DOI: 10.1201/9781003356172-5

FIGURE 4.1 Operators: a way to request information. [Image from www.flaticon.com/free-icon/call-center-operator_49128]

4.2 OSCILLATING PROBABILITY

The wave function $\Psi = e^{i(kx-\omega t)}$ for a free particle is oscillating. If the particle is free, why does it have varying probability to be at different positions at different times?

Answer: it does *not*! Ψ varies with x and t, but $|\Psi|^2$ does not (problem 4.2). The real and imaginary parts of $e^{i(kx-\omega t)} = \cos(kx-\omega t) + i\sin(kx-\omega t)$ oscillate 90° out of phase with each other such that their vectorially combined magnitude is constant. The electron is equally likely to be found at any position x, at any time t. It is free.

4.3 UNCERTAINTY

For the free particle wave function $\Psi = e^{i(kx-\omega t)}$, p is known exactly ($p = \hbar k$). But what is known about position? As just discussed, the particle is equally likely to be found everywhere along x. This is a consequence, indeed a partial proof, of the Heisenberg uncertainty principle:

$$\Delta p\, \Delta x \geq \hbar/2 \quad \text{[Heisenberg uncertainty principle]} \tag{4.4}$$

In words: the uncertainty in momentum multiplied by uncertainty in position is greater than or equal to Planck's constant (over 4π). Since $\hbar$ is a small number, the product of these uncertainties is small, but the key point is that the product can *not* be zero.

FIGURE 4.2 Heisenberg uncertainty principle: position and momentum can not be simultaneously known.

This equation dictates that in the case of a free particle, where the momentum is known exactly ($\Delta p \rightarrow 0$), the position must not be known at all ($\Delta x \rightarrow \infty$). Analogously, if we wish to localize a particle to a specific position in space x such that $\Delta x \rightarrow 0$, the particle's momentum must become totally uncertain ($\Delta p \rightarrow \infty$). Thus, an infinite number of waves with different values of k would have to be linearly combined in order to localize the particle. This concept of a "wave packet" can be used to prove the Heisenberg uncertainty principle.

4.4 QUANTA

The word "quantum" means a discreet amount of energy. The term "quantum mechanics" would seem to imply that an electron can only have certain energies. However, a free electron can have any energy ranging from zero to infinity (in theory), since there is no restriction on the value of k in its wave function $\Psi = e^{i(kx-\omega t)}$. This does not seem like quantum mechanics: the allowed energy levels for the free electron are not "quantized". But in some circumstances, the energy is indeed quantized. For instance, if the electron is confined to a box (Chapter 8), or if the electron is confined to be near a nucleus (part of an atom), then it can only have certain energies. If the electron breaks out of the box, or away from the atom, it can then have any of a number of continuous energies, because it is free.

The circumstances in which the electron is trapped lead to solutions that are called "bound states". For these bound states, energies are quantized. It is only because bound (quantized) states are so important for understanding atoms and molecules, and lead to interesting conclusions about atomic and molecular behavior, that this topic is called quantum mechanics.

PROBLEMS

4.1 **(a)** Assume a free particle wave function $\Psi = e^{i(kx-\omega t)}$. Use the operators in equation 4.1 to find the energy and the momentum of the particle. **(b)** How do these answers change if the particle's wave function is instead $\Psi = e^{i(-kx-\omega t)}$? **(c)** Based on your answers from parts (a) and (b), what is the difference between the two particles?

4.2 Show how the probability function P(x,t) for the free particle wave function $\Psi = e^{i(kx-\omega t)}$ varies with x and t. What does this tell you about where and when the particle is found?

4.3 If a one-dimensional free particle is known to be present only between positions x = 1.2 meters and x = 1.6 meters, then how precisely can its momentum be known?

4.4 Write a brief paragraph about wave packets after reading about wave packets from another source.

5 Separation of Variables

5.1 WHAT IS "SEPARATION OF VARIABLES"?

As introduced in Chapter 4, the one-dimensional Schrödinger equation (TDSE) is:

$$\mathcal{H}\Psi = i\hbar\,(\partial/\partial t)\Psi \tag{5.1}$$

where $\mathcal{H}$ is called the Hamiltonian operator and represents $[-\hbar^2/2m(\partial/\partial x)^2 + V]$. Note that $\Psi(x,t)$ is a function of both x and t. In this chapter, the TDSE will be separated into two equations: one for x and one for t. This will also produce two separable parts of Ψ so that $\Psi(x,t)$ can be written as a function of x times a function of t. After this separation, solving the Schrödinger equation will be much simpler.

5.2 IN ORDER TO SEPARATE SPACE FROM TIME, THE POTENTIAL MUST BE TIME-INDEPENDENT

We have discussed that both sides of the TDSE are related to energy: $\mathcal{H}$ and $i\hbar(\partial/\partial t)$ both can be thought of as energy operators. But they "extract" the energy from Ψ by different means. $\mathcal{H}$ uses x derivatives, while E uses a t derivative. Another way of stating this is that $\mathcal{H}$ involves only space and E involves only time. Of course, this is only true if the potential V is *time-independent*. In this chapter, we will always assume V is time-independent (the potential varies only with position), so that $V = V(x)$. In that case, special *separable* solutions Ψ can be found that solve the TDSE. By separable, we mean solutions wherein the x and t parts of Ψ can be separated and multiplied by each other:

$$\text{separable solution: } \Psi(x,t) = \psi(x)f(t) \; [\text{e.g. } \Psi = e^{i(kx-\omega t)} = (e^{ikx})(e^{-i\omega t})] \tag{5.2}$$

An example of a function that is *not* separable is $\Psi = \sin(kx-\omega t)$.

5.3 PLUG THE SEPARABLE FORM OF Ψ (x,t) INTO THE TDSE

If we assume a separable solution, the TDSE can be simplified by plugging the separable form of Ψ (equation 5.2) into the TDSE (equation 5.1):

DOI: 10.1201/9781003356172-6

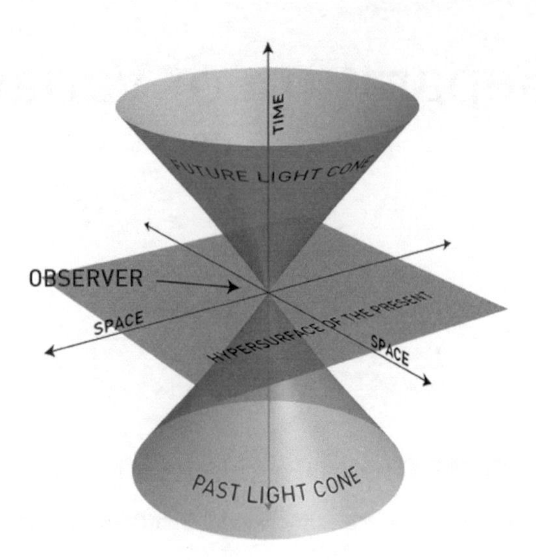

FIGURE 5.1 Space-time cone. Space and time are related, but we must separate them to derive the time-independent Schrödinger equation (TISE). [Image by © Stib, https://en.wikipedia.org/wiki/File:World_line.png, licensed under CC BY-SA 3.0]

$$\mathcal{H}\psi(x)f(t) = i\hbar(\partial/\partial t)\,\psi(x)f(t) \quad (5.3)$$

Since f(t) has no x dependence, it can pass through the $\mathcal{H}$ operator. Similarly, since $\psi(x)$ has no t dependence, it can pass through $(\partial/\partial t)$:

$$f(t)\,\mathcal{H}\psi(x) = \psi(x)\,i\hbar(\partial/\partial t)f(t) \quad (5.4)$$

This does not seem like a simplification. To simplify, divide both sides by $\Psi = \psi(x)f(t)$:

$$f(t)\,\mathcal{H}\psi(x) / \psi(x)f(t) = \psi(x)\,i\hbar(\partial/\partial t)f(t) / \psi(x)f(t) \quad (5.5)$$

Next, cancel numerator and denominator terms not acted on by operators. The LHS has f(t)/f(t) = 1, the RHS has $\psi(x)/\psi(x) = 1$. These cancelations produce:

$$\mathcal{H}\psi(x) / \psi(x) = i\hbar\,(\partial/\partial t)f(t) / f(t) \quad (5.6)$$

No further cancelations can be made since the functions in the numerators are subject to derivative-containing operators, which can change the form of the function they operate upon.

5.4 SEPARATE INTO THE TISE AND SISE

The key concept in the separation procedure is the realization that the LHS in equation 5.6 involves only x, while the RHS involves only t. This is intriguing. We now have two expressions – the LHS being one, and the RHS being the other – that are always equal. And yet, if the value of only x is changed, clearly the RHS does not change. This means that the LHS cannot change either. The LHS must thus be a constant. And since LHS = RHS, then the RHS must equal the same constant. Let's call that constant E (yes, it is energy). Equation 5.6 can now be split into two equations:

$$\mathcal{H}\,\psi(x)\,/\,\psi(x) = E, \quad i\hbar\,(\partial/\partial t)\,f(t)\,/\,f(t) = E \tag{5.7}$$

These are rearranged to:

$$\underset{\text{[TISE]}}{\mathcal{H}\,\psi(x) = E\,\psi(x),} \qquad \underset{\text{[SISE]}}{i\hbar\,(\partial/\partial t)\,f(t) = E\,f(t)} \tag{5.8}$$

These two equations arise from the TDSE, but the left equation involves only space, while the right equation involves only time. Therefore, these equations can be called the time-independent Schrödinger equation (TISE) and the space-independent Schrödinger equation (SISE)[1].

From Chapter 4, it should be clear that the TISE and SISE are two eigenvalue equations, where $\psi(x)$ is an eigenfunction of $\mathcal{H}$ with eigenvalue E, and f(t) is an eigenfunction of the $\hat{E}$ operator, also with eigenvalue E. You see now why we chose E as the separation constant above.

5.5 SEPARABLE SOLUTION ↔ SINGLE ENERGY ↔ EIGENFUNCTION

To summarize: if the potential is time-independent, then it is possible to find separable wave functions [$\Psi = \psi(x)f(t)$] that solve the TDSE. Not all possible solutions will be separable, but it is possible to find some that are. These separable solutions are very important because, according to equations 5.8, their parts $\psi(x)$ and f(t) are eigenfunctions of the $\mathcal{H}$ and $\hat{E}$ operators. That is, they have well-defined energy. That means the energy levels of the electrons in an atom (ε_{1s}, ε_{2s}, ε_{2p}, ε_{3s} … .), which are well-defined, must correspond to separable solutions to the TDSE. In fact, we can find the energies of each level of a hydrogen atom by solving the eigenvalue equation $\mathcal{H}\psi = E\psi$ using the potential V that is appropriate for hydrogen. There will be distinct separable solutions corresponding to the 1s, the 2p, the 3d orbitals, etc.

5.6 SOLVING THE SISE (ONCE)

Setting aside the TISE, we first solve the SISE:

$$\text{SISE:} \quad i\hbar\,(\partial/\partial t)\,f = E\,f \tag{5.9}$$

To simplify, move all constants to the RHS:

$$(\partial/\partial t)\,f = (E/i\hbar)\,f = (\omega/i)\,f = (-i\omega)\,f \tag{5.10}$$

where $E = \hbar\omega$ and $1/i = -i$ were used. By equating the first and last expressions in equation 5.10, it becomes clear that we seek a function f(t) whose first derivative produces the same function back again, times $-i\omega$. The function $f(t) = e^{-i\omega t}$ works perfectly.

Note that the SISE never changes: the potential V appears only in the TISE. We have solved the SISE for *all* cases where V is time-independent, whether the particle is free or trapped. The solution is *always* $f(t) = e^{-i\omega t}$ and indicates a particle with energy $E = \hbar\omega$, as can be shown by plugging f(t) into the SISE (equation 5.9).

5.7 SOLVING THE TISE (OVER AND OVER)

Note that the TISE (see equation 5.8) *does* depend on the form of V(x): the TISE cannot be solved until V(x) is specified. Therefore, whether the electron is free or trapped and, if the latter, the nature of the trap *do* affect $\psi(x)$. There is no "one size fits all" solution for the TISE, as there is for the SISE.

The principal task for the next few chapters is to solve the TISE (find $\psi(x)$) for various simple potentials V. The full wave functions $\Psi(x,t)$ can then be formed by multiplying each eigenfunction $\psi(x)$ by $e^{-i\omega t}$. In practice, we need not very often explicitly multiply by $e^{-i\omega t}$. The spatial eigenfunctions $\psi(x)$ often suffice to specify the energy, position, momentum, etc., of the electron.

5.8 SUPERPOSITION OF EIGENFUNCTIONS

The *superposition principle* states that any linear combination of efs Ψ_n is also a valid solution of the TDSE. In that case, Ψ will not be separable, but it can be written as a superposition of two or more distinct eigenfunctions $\psi(x)$, each multiplied by its own distinct $e^{-i\omega t}$ factor. The equation $E = \hbar\omega$ tells us that two or more distinct values of E will thus be involved, and that the electron therefore does not have a well-defined energy. An example of the superposition principle is discussed in Chapter 9.

PROBLEMS

5.1 The LHS of equation 5.6 depends on $\psi(x)$, and $\psi(x)$ can be a quite complicated function. Since this is the case, discuss how it could be that the LHS is actually equal to a constant (called E) as shown in equation 5.7.

5.2 Show that any separable solution to the TDSE, which *must* always take the form $\psi(x)\, e^{-i\omega t}$, has **(a)** no time dependence in its probability function *P(x,t)* and **(b)** a fixed energy. **(c)** What is the value of the fixed energy?

5.3 In Section 5.1, $\Psi = \sin(kx-\omega t)$ was given as an example of a *non*-separable solution. For this form of Ψ, show: **(a)** that $\Psi = \sin(kx-\omega t)$ is *not* a solution to the TDSE for a free particle, **(b)** the value of the kinetic energy KE, and **(c)** that the probability function *P(x,t)* is *not* constant.

NOTE

1 TISE is common usage, but most texts do not use the term SISE. They could.

6 $\psi(x)$: General Conditions, Normalization, Bra-kets

6.1 GENERAL CONDITIONS FOR $\psi(x)$

In the next chapter, the TISE will be solved for a few simple potentials. But first, we discuss the types of functions $\psi(x)$ that are acceptable. We may find a solution $\psi(x)$ that mathematically solves the TISE, but does not make sense physically. For instance, perhaps a rising exponential mathematically solves the TISE for some $V(x)$, but this solution may be unacceptable since, as $x \to \infty$, the probability also approaches infinity. The general conditions required for ψ and its derivative or slope $(\partial/\partial x)\psi$, the latter of which we write as ψ', are as follows:

ψ and ψ' must each be single valued, finite, and continuous

Single-valued: We could envision a function that has two values at, say, $x = 3$. This is clearly nonsense, as there cannot be two probabilities to find the particle at $x = 3$. If ψ is single-valued, then clearly ψ' must also be single-valued.

Finite: As discussed above, $\psi(x)$ cannot go to infinity, since the probability would also go to infinity. However, a rising exponential may be acceptable if it only applies over a finite area. We will shortly encounter decaying exponentials that are also acceptable. Also, ψ' cannot be infinite at any point: if it is, then the value of ψ itself must be infinite at that point. In fact an infinite slope also implies that the function is multi-valued at that point.

Continuous: The requirement for continuity of ψ and ψ' follows similar arguments. Continuity means that the function cannot skip from one value to a distant value when moving along x by an infinitesimal amount. If $\psi(x)$ does change in that way, then the slope is infinite, violating the previous condition. The slope must also be continuous. This simply means that there can be no perfectly sharp corners in $\psi(x)$. Why? To understand, note that a discontinuous slope requires an infinite second derivative. The second derivative $(\partial/\partial x)^2$ appears in the Hamiltonian. Therefore, an infinite second derivative is associated with infinite KE. The electron cannot exist at that point unless it has infinite energy. Realistically, then, the electron cannot exist at that point.

DOI: 10.1201/9781003356172-7

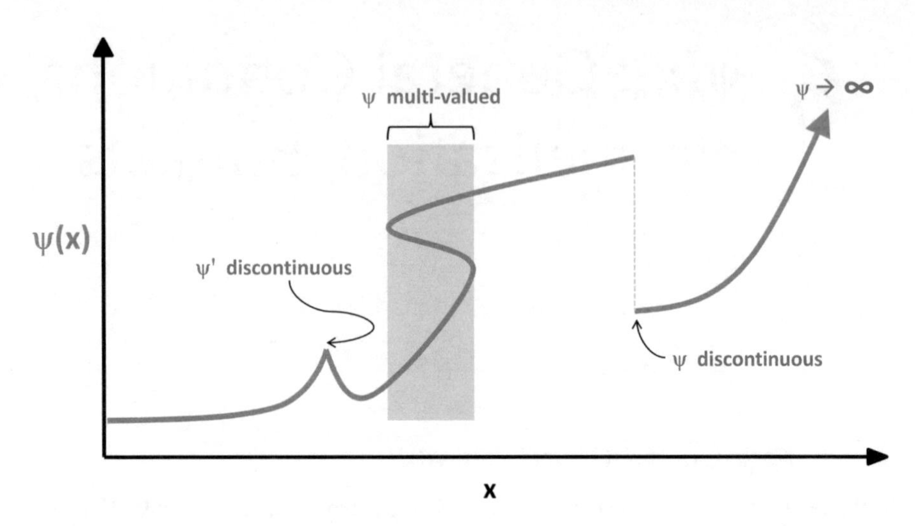

FIGURE 6.1 Unacceptable wave functions. ψ and ψ' must be single-valued, finite and continuous.

This last explanation of why ψ' cannot be discontinuous points out the possibility of an exception:

Exception: ψ' may be discontinuous at an infinite discontinuity of the potential

If the particle cannot exist at a particular point x, then the slope can be discontinuous at that point. When can the electron not be at a particular point? When $V \rightarrow \infty$. Therefore, if there is a discontinuity in the potential V, such that V is finite at some position x and then V increases to infinity at the adjacent position, then $\psi(x)$ may have a sharp corner at that point (see PIB discussion in Chapter 8).

6.2 NORMALIZATION OF $\psi(x)$

Normalization simply means scaling the wave function so that the total probability to find the particle anywhere in space equals 1. Since $P(x) = \psi^*(x)\psi(x)$, then the total probability for the particle to be found somewhere along x is:

$$P(\text{to be some where along } x) = \int_{-\infty}^{+\infty} dx\, \psi^*(x)\psi(x) \tag{6.1}$$

If this integral does not equal 1, then $\psi(x)$ is not of the right size: $\psi(x)$ should be multiplied by the appropriate correction factor to make the integral equal to 1. This factor is called the *normalization constant*. It is not always necessary to normalize a wave function in order to extract information. An example of the normalization procedure is presented in Chapter 8.

"A theory with mathematical beauty is more likely to be correct than an ugly one that fits some experimental data."

FIGURE 6.2 Paul Dirac introduced bra-ket notation and developed relativistic QM for fast particles.

[Photo from https://en.wikipedia.org/wiki/Paul_Dirac#/media/File:Paul_Dirac,_1933.jpg, public domain]

6.3 DIRAC BRA-KET NOTATION

Integration of $\psi^*\psi$ over all values of x is such a common operation in QM that a special shorthand notation has been developed:

$$\int_{-\infty}^{+\infty} dx\, \psi^*(x)\psi(x) = <\psi|\psi> \tag{6.2}$$

The symbol $<\psi|\psi>$ is called a Dirac bra-ket, and is composed of a bra $<\psi|$ and a ket $|\psi>$. The bra $<\psi|$ implicitly contains the complex conjugate of ψ. When the bra and ket are combined, then and only then is integration performed. For instance, an eigenstate, or any state, can be written as $|\psi>$, and the fact that the state is normalized is written as $<\psi|\psi> = 1$. Bra-ket notation is useful shorthand when integrals are to be performed.

PROBLEMS

6.1 Draw six one-dimensional wave functions to illustrate violation of each of the six general conditions in Section 6.1.

6.2 Suppose you find that $\langle\psi|\psi\rangle = 38$. What is the normalization factor for ψ?

6.3 Write the TISE in bra-ket notation. Do the same for the TDSE. (Hint: use only kets)

6.4 Starting from the ket form of the TISE (see Problem 6.3), operate on both sides with the bra form of ψ. Simplify. Attempt to interpret the meaning of the resulting equation.

One-Dimensional Potentials

7 Solving the TISE for the Simplest Potentials

In order to show how the TISE can be used to find eigenfunctions, a few simple potentials will be introduced in this chapter. The emphasis is on visually understanding the shape that the wave function is forced to take. Some calculations will be necessary, but your focus should be on the visual results.

7.1 FREE PARTICLE

For a free particle (V = 0), the TISE is:

$$\mathcal{H}\psi = E\psi \rightarrow [-\hbar^2/2m(\partial/\partial x)^2]\psi = E\psi \tag{7.1}$$

We guess at a solution $\psi = e^{+ikx}$. Plugging in confirms that it is a solution, provided that $\hbar^2k^2/2m = E$.

Note that $\psi = e^{+ikx}$ is a complex function that is simply shorthand notation for a real wave [cos(ωt)] plus an imaginary wave [*i*sin(ωt)]. Quantum texts typically ignore this fact and draw just a single sinusoid-shaped wave to represent $\psi = e^{+ikx}$. We will follow suit (see figure 7.1). This simple representation suffices to show wavelength (related to energy) and amplitude (related to probability).

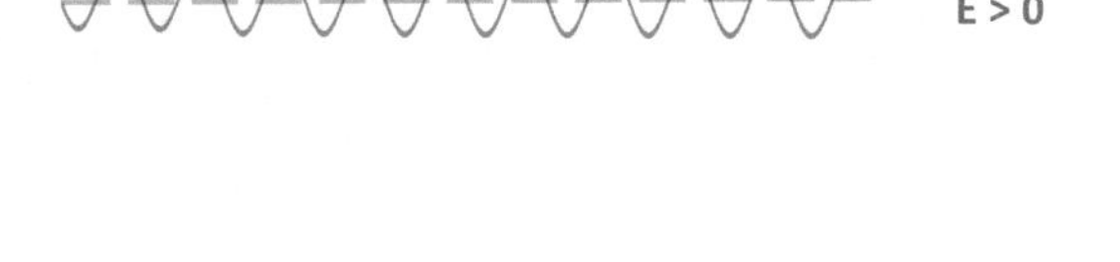

FIGURE 7.1 A free particle wave function. The potential (V = 0) is drawn in black, the energy level as a dashed grey line, and the wave function $\psi = e^{+ikx}$, oscillates about the dashed line. The wave function is drawn as a simple sinusoidal function, which suffices to indicate the energy and probability.

DOI: 10.1201/9781003356172-9

Note that $\psi = e^{-ikx}$ is also a solution to equation 7.1. What is the difference between these two possible solutions? We can "interrogate" these two functions via the momentum operator:

$$\hat{p}\, e^{+ikx} = \hbar/i\, (\partial/\partial x)\, e^{+ikx} = \hbar/i\, (+ik)\psi = +\hbar k\, \psi \rightarrow p = +\hbar k \qquad (7.2)$$

$$\hat{p}\, e^{-ikx} = \hbar/i\, (\partial/\partial x)\, e^{-ikx} = \hbar/i\, (-ik)\psi = -\hbar k\, \psi \rightarrow p = -\hbar k \qquad (7.3)$$

Apparently, the two wave functions ($e^{\pm ikx}$) represent the same magnitude of momentum ($\hbar k$) but in opposite directions, along +x or –x, respectively. Since the magnitude of momentum is the same, the KE is the same ($KE = p^2/2m$). These two functions can be combined to form a single function:

$$\rightarrow \psi = Ae^{+ikx} + Be^{-ikx}$$

(where A and B indicate the probability of + vs. – momentum) (7.4)

Although this combined function is *not* an eigenfunction of the $\hat{p}$ operator, it *is* a solution to the TISE, and hence it *is* an eigenfunction of the $\mathcal{H}$ operator (see problem 7.1. Problem 7.2 discusses the related $\psi(x,t)$ function). No restrictions arise for the possible values of k, and therefore any non-negative energy is allowed for a free particle. This concept is related to continuum energies encountered in certain types of spectroscopy, when the electron receives enough energy to break *free* of the atom or molecule.

7.2 CONSTANT POTENTIAL: $V = V_0$ EVERYWHERE

Suppose the particle is a conduction electron in a large block of metal, free to move anywhere within the metal. Although its movement is free, its potential energy V is not zero, because it would take energy to remove it from the metal. Therefore, the potential V is constant at a value that we will call V_0. How does this affect the solutions to the TISE and the resulting energies? The 1D TISE becomes:

$$[-\hbar^2/2m(\partial/\partial x)^2 + V_0]\psi = E\psi \qquad (7.5)$$

It can be guessed that solutions of the form $\psi = e^{ikx}$ or $\psi = e^{-ikx}$ solve this equation, as they did for the TISE in free space. However, on plugging either one of these solutions into equation 7.5, (see problem 7.3) we find that the equation is satisfied only if:

$$\hbar^2 k^2/2m + V_0 = E \quad \text{or} \quad k = 1/\hbar\sqrt{2m(E - V_0)} \qquad (7.6)$$

Thus, while the solutions look the same as for a free particle, the relationship between k and E has been altered by V_0. This can be understood by recalling that k is related to momentum and therefore to KE. Equation 7.6 simply states that KE + V = E. The solutions are thus moving waves. But for a given E, k and momentum are less than

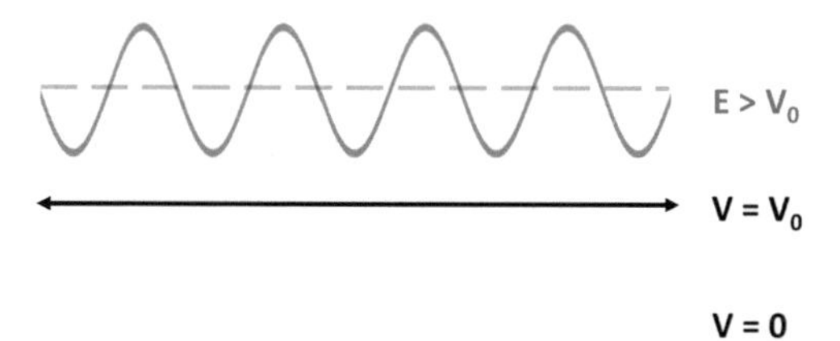

FIGURE 7.2 A particle subject to a constant potential V_0. The particle energy is the same as in Figure 7.1. The particle moves more slowly then in Figure 7.1, because some of its energy is potential energy (V_0), which reduces the available KE. Therefore, the wave length is longer. Also, the amplitude is higher, which can be rationalized by noting that a slower moving particle will spend more time traversing a given region.

for a free particle, since some of the energy is tied up as potential energy (V_0). The wavelength, inversely related to k, is thus longer (see figure 7.2).

As for a free particle, this electron wave can travel in either direction, and so the general solution is of the form:

$$\psi = Ae^{+ikx} + Be^{-ikx} \tag{7.7}$$

Any value of $E \geq V_0$ is allowed. Note that the discussion and equations in this section reduce to those in section 1 if V_0 is set to zero (problem 7.4).

7.3 POTENTIAL STEPS, BARRIERS, AND WELLS

The remaining potentials in this chapter can be treated as was the constant potential in section 7.2. The only difference is that there will now be two or three regions, each with a distinct constant value of the potential. Therefore, *distinct wave functions must be defined for each region*. In addition, as discussed in Chapter 6, these distinct wave functions and their slopes must be *continuous* with each other at the points where they meet (at the *boundaries* where V changes).

This type of problem is called a *boundary value* problem. The process of making the function and its derivative continuous across the boundaries is called matching the *boundary conditions* (the conditions are continuity). In the process, information regarding the likelihood that the particle can traverse the boundary will arise, leading to the concept of *transmission coefficient* (T), *reflection coefficient* (R), and *tunneling*. In the next chapter, we will see that if the potential traps the particle in a finite region of space, the boundary conditions will lead to quantization of the energy levels.

7.4 FINITE POTENTIAL STEP: CASE OF $E > V_0$

Suppose that the conduction electron in the previous section encounters the edge of the metal. It is simpler mathematically to assume that the potential within the metal is zero, and that as it leaves the metal at x = 0, it moves to a positive constant potential environment ($V = V_0$). This is equivalent to a ball rolling along the ground and

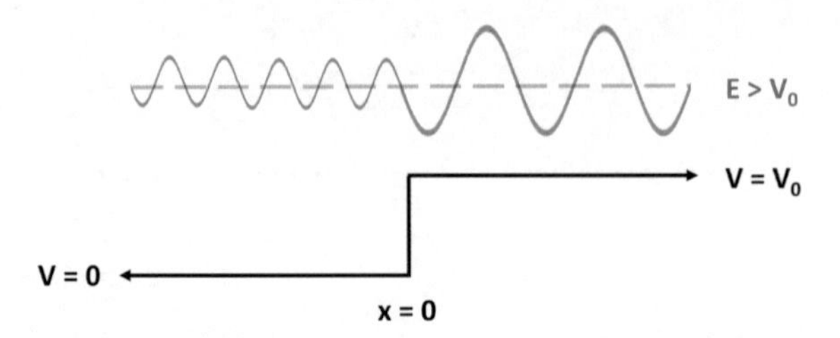

FIGURE 7.3 A finite potential step with $E > V_0$. The potential is drawn in black, the energy level as a dashed grey line, and the wave function oscillates about the dashed line. The particle is assumed to be incident from the left. Note the similarities to Figures 7.1 and 7.2. Although the particle has sufficient energy to overcome the potential barrier at x=0, there is some probability that it will be reflected by the barrier, as calculated in the text.

then encountering a ramp up onto a table top: On the floor, all of its energy is kinetic. As the ball rolls up the ramp, some of the KE converts to potential energy (PE), and the ball slows down. In QM, momentum is used rather than velocity: past $x = 0$, the momentum reduces and the wavelength increases, since $k = 2\pi/\lambda$ (see figure 7.3). In the classical analogy, the total energy of the ball is the same on the floor as on the table. Similarly, the total E of the electron does not change as it leaves the metal, and so via $E = \hbar\omega$, the value of ω does not change.

The ball analogy helps explain roughly what we expect to happen to the electron. Next, we seek the eigenfunctions for an electron encountering the barrier. We need distinct eigenfunctions for the two regions of space. We already know the appropriate eigenfunctions from sections 1 and 2:

$$x < 0: \quad \psi = Ae^{+ikx} + Be^{-ikx}, \quad \text{where } k = 1/\hbar\sqrt{2m(E)} \tag{7.8}$$

$$x > 0: \psi = Ce^{+ik'x} + De^{-ik'x}, \quad \text{where } k' = 1/\hbar\sqrt{2m(E - V_0)} \tag{7.9}$$

where the symbol k' is introduced for k outside the metal. These two wave functions suggest that the electron can move forward or backward in each region, but will move more slowly in the $x > 0$ region, since $k' < k$ for a given value of E.

The constants A, B, C, and D are not known at this point. They indicate the relative probabilities that the particle has forward vs. backward momentum, and that it is found inside vs. outside the metal. To find the constants, we first simplify by defining the problem as an electron incident from the left (the A term above). It encounters the barrier (the edge of the metal) and has some probability to bounce back into the metal (negative momentum: the B term in eq. 7.8) or to travel forward past $x = 0$ (the C term in eq. 7.9). There is no reason to assume the particle, after leaving the metal and travelling through free space for some distance, will turn around and head back toward the metal, so there is no need for the "D" term, and we set $D = 0$.

The values of B and C must therefore be related to the probability of a particle, on encountering the barrier at $x = 0$, being reflected (B) or transmitted (C). To analyze

FIGURE 7.4 Reflection and Transmission. Light can be transmitted (foreground) or reflected (background) at the boundary between air and water. [Photo by © Scott McGuire, https://fineartamerica.com/featured/tenaya-lake-yosemite-national-park-scott-mcguire.html, by permission]

these probabilities, two general conditions for wave functions from Chapter 8 are used: ψ and ψ' must be continuous. The functions in equations 7.8 and 7.9 are continuous within their own regions. They should also be continuous with each other when they meet at the point x = 0. Also, their first derivatives should be equal at x = 0:

$$\psi(0) = Ae^{ik0} + Be^{-ik0} = Ce^{ik'0} \rightarrow A + B = C \tag{7.10}$$

$$\psi'(0) = ikAe^{ik0} + -ikBe^{-ik0} = ik'Ce^{ik'0} \rightarrow A - B = (k'/k)C \tag{7.11}$$

Addition and subtraction of these two equations on the right, and division by two, yields:

$$A = (C/2)(1 + k'/k); \quad B = (C/2)(1 - k'/k) \tag{7.12}$$

Using these two equations, A and B can be eliminated (in favor of C) from the eigenfunctions of equations 7.8 and 7.9. Although C is not known, the relative sizes of A vs. B vs. C are known. Since probability goes as $|\psi|^2$, the probability of the particle bouncing back off the barrier (R) can be derived via ratios comparing B^2 to A^2:

$$R = |B/A|^2 = |(1 - k'/k)/(1 + k'/k)|^2 \quad R = \text{Reflection coefficient} \tag{7.13}$$

A simple rearrangement yields (problem 7.7):

$$R = |(k - k')/(k + k')|^2 \tag{7.14}$$

Since the total probability should be 1, the probability of transmission through the barrier can be simply calculated:

$$T = 1 - R = 1 - |(k-k')/(k + k')|^2 \tag{7.15}$$

Equation 7.15 can be simplified to yield (problem 7.8):

$$T = 4kk'/(k + k')^2 \tag{7.16}$$

We won't dwell on the equations for R and T. We will merely point out that the probabilities of transmission (leaving the metal) or reflecting back into the metal depend only on the values of k and k′; that is, the momentum inside and outside the metal. For instance, if the barrier is very small, then k and k′ are almost equal, and therefore R is nearly zero: little chance of reflection from a tiny barrier.

Interestingly, although we assume in this section that $E > V_0$, there is still some probability of reflection at the boundary. This certainly differs from the classical analogy of the ball rolling up the ramp: there is no chance of the ball failing to roll all of the way up the hill, so long as $E > V_0$. Reflection is therefore clearly a quantum phenomenon, as is the associated transmission.

7.5 FINITE POTENTIAL STEP: CASE OF $E < V_0$ (TUNNELING)

Classically, if a ball has not enough KE to roll up the ramp, then there is no probability to find it on top of the table. This is not so in quantum mechanics. The mathematics for this situation follows the same logic as in the section 7.4, except that k′, which involves the square root of $(E\text{-}V_0)$, is imaginary. The function $e^{ik'x}$ thus becomes a real exponential, decaying as x increases. To simplify analysis, it is customary to define a new variable κ:

$$\kappa = 1/\hbar\sqrt{2m\left(V_0 - E\right)};\quad \kappa = k'/i \tag{7.17}$$

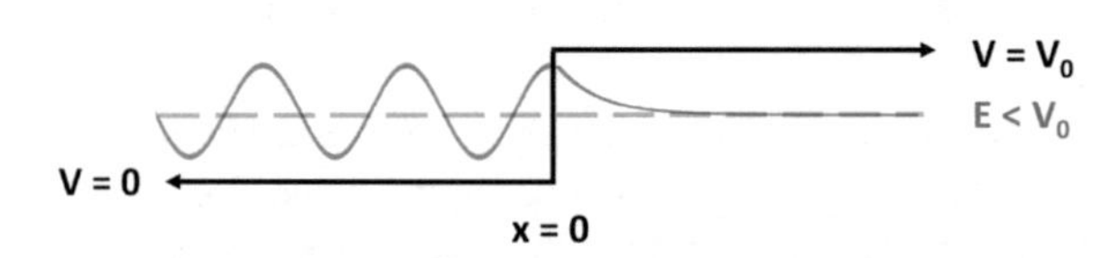

FIGURE 7.5 A finite potential step with $E < V_0$. The particle is assumed to be incident from the left. Although the particle lacks sufficient energy to overcome the barrier, there is some probability that it will be found within the barrier region.

The treatment follows that of the section 7.4, with the wave function for x > 0 taking the form:

$$x > 0: \quad \psi = Ce^{+ik'x} \rightarrow C\,e^{-\kappa x} \tag{7.18}$$

From the shape of this function, it is apparent that the electron *can* enter the barrier, even though it would theoretically have negative KE at that position (since $E < V_0$). Very strange indeed. This phenomenon, of a particle finding its way into a classically forbidden region, is called *tunneling* "into" a barrier. The probability of finding the electron in the barrier reduces exponentially with depth x. Tunneling also reduces with increased κ, that is, for large $(V_0 - E)$.

7.6 FINITE BARRIER OF FINITE WIDTH: CASE $E > V_0$

Suppose the electron approaches a barrier from the left, but now assume that the barrier only extends a short distance. What is the probability that the electron will traverse the barrier?

This can be calculated (we will not show the results) by the same approach as in the previous section, with three distinct wave functions for the three regions, and matching of conditions at the two boundaries. Since we assume $E > V_0$ in this section, Ψ is oscillatory in each region, although wavelength is longer within the barrier. What will be the KE of the particle on the right side of the barrier? (Think of the ball.)

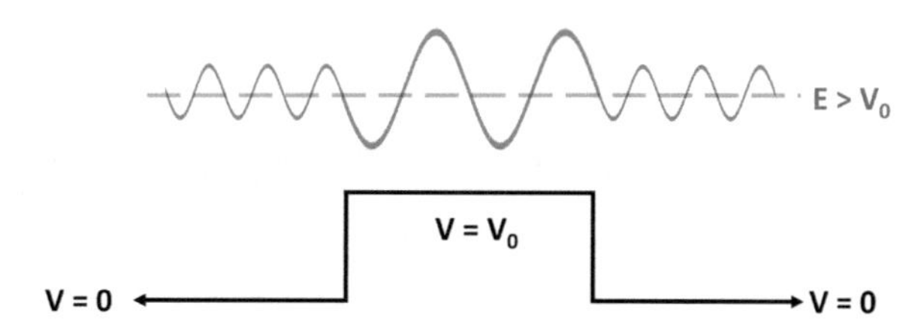

FIGURE 7.6 A finite rectangle barrier of height V_0 with $E > V_0$. Note the similarity to Figure 7.3.

7.7 FINITE BARRIER OF FINITE WIDTH: CASE $E < V_0$

If the electron energy is less than V_0, then classically the electron cannot enter the barrier. However, we know from section 7.5 that a QM electron *can* enter a barrier that is higher than its energy. In this case, the wave function would be oscillatory on either side of the barrier, but a decaying exponential within the barrier, of the form $\psi = e^{-\kappa x}$. We are interested in the probability that the electron incident from the left will tunnel *through* the barrier (make it to the other side).

Calculation of the probability is relatively simple. ψ(0) indicates the probability of finding the electron at the start of the barrier, and ψ(L) indicates the probability finding the electron at the end of a barrier of length L. Therefore, the ratio of the two

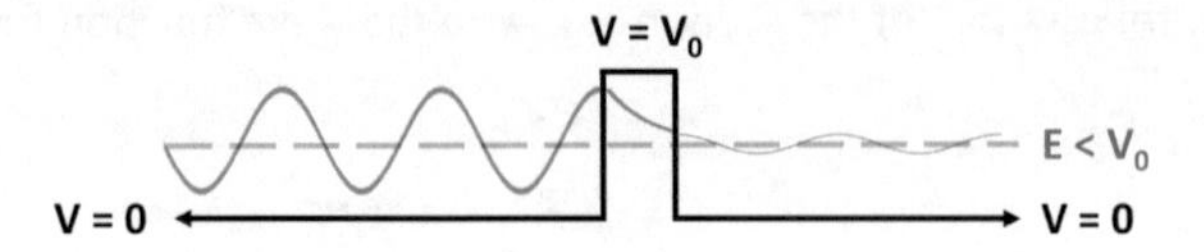

FIGURE 7.7 A finite rectangle barrier of height V_0 with $E < V_0$. Note the similarity to Figure 7.5. The particle, incident from the left, has some probability to *tunnel* through the barrier, although classically it would have insufficient energy to overcome the barrier. Tunneling probability can be calculated from the rate of exponential decay within the barrier. Note that the wavelength is the same to the left of and to the right of the barrier region.

probabilities provides the probability that a given electron, incident from left, will tunnel through the barrier:

$$T = \left(e^{-\kappa L}\right)^2 / \left(e^{-\kappa 0}\right)^2 = \left(e^{-\kappa L}\right)^2 / 1 = \left(e^{-2\kappa L}\right) \tag{7.19}$$

Each term is squared to turn the wave function amplitude information into probability information. Apparently, the probability of tunneling depends only on the value of κ and the width of the barrier L. Equations 7.17 and 7.19 together indicate that higher and longer barriers reduce the probability of tunneling, though more energetic electrons have a better chance. This sort of very simple argument can be used to calculate the probability that a nucleus will decay, via for instance an alpha particle tunneling through the nuclear potential (see problem 7.10).

7.8 FINITE BARRIER OF FINITE WIDTH: CASE $E < V_0$ (FINITE WELL)

If the potential from figure 7.7 is inverted, the barrier becomes a well. A particle with energy E less than the height of the well walls (V_0) will be trapped within the well, sort of. The wave function will oscillate within the well, but it will decay exponentially outside of the well. In the next chapter, we will start from this point, but assume that the potential walls are infinitely high.

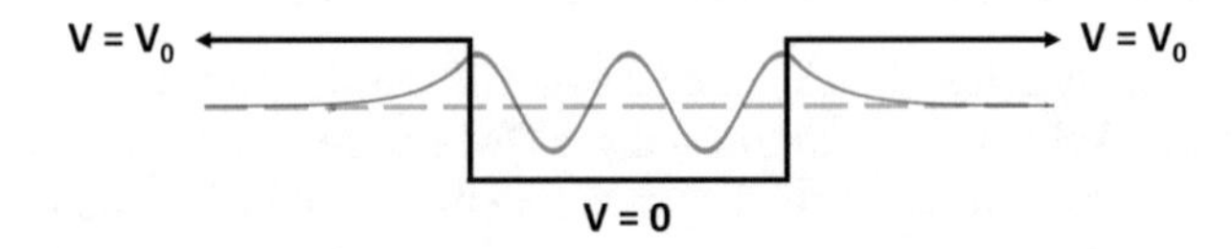

FIGURE 7.8 A finite rectangular well with walls of height V_0 and with $E < V_0$. The solution oscillates within the well, where V = 0. Within the walls of the well, since $E < V_0$, the solution is a decaying exponential: the particle has some probability to be found outside of the well.

PROBLEMS

7.1 For the free particle wavefunction $\psi = Ae^{ikx} + Be^{-ikx}$, show whether ψ is an eigenfunction of **(a)** the Hamiltonian; **(b)** the momentum operator. If it is an eigenfunction, determine the eigenvalue.

7.2 **(a)** Starting from the free particle wave function $\psi(x) = Ae^{ikx} + Be^{-ikx}$, create the time-dependent wave function $\Psi(x,t)$. **(b)** Calculate the probability function for this form of $\Psi(x,t)$. **(c)** given this probability function, is the function $\Psi(x,t)$ a stationary state? Is $\Psi(x,t)$ an eigenfunction of $\mathcal{H}$?

7.3 Plug $\psi = Ae^{ikx}$ or $\psi = Be^{-ikx}$ into equation 7.5 and prove equation 7.6.

7.4 Show that if $V_0 = 0$, equations 7.6 revert back to their forms in section 7.1 (free particle).

7.5 Suppose the constant potential V_0 in section 7.2 was negative instead of positive. In this case, which would be greater, KE or E? Explain.

7.6 Why does the D term not appear in equation 7.10?

7.7 Prove equation 7.14, starting from equation 7.13.

7.8 Prove equation 7.16 starting from equation 7.15. For simplicity, assume k and k′ are real.

7.9 Evaluate R and T when **(a)** $V_0 \rightarrow 0$; **(b)** $V_0 \rightarrow E$. Explain the meaning of your results.

7.10 Decay/escape of an alpha particle from the nucleus is a tunneling phenomenon. The alpha particle is trapped in the nucleus, but has a certain probability to tunnel through the nuclear barrier. Calculate this tunneling probability for one approach of the alpha particle. Assume the following: alpha particle E = 10 MeV, rectangular barrier of height 20 MeV, and width 30 femtometers.

8 The One-Dimensional Particle in a Box

The particle in a box (PIB), also known as the infinite square well, is an extremely useful quantum mechanical system: the eigenfunctions are visually and mathematically simple, and easy to obtain. The PIB is thus ideal to illustrate the principles of QM, including stationary states, normalization, orthogonality, expectation values, and the superposition principle.

8.1 FIND THE PIB EIGENFUNCTIONS

Suppose $V = 0$ for $0 < x < L$ and $V \to \infty$ outside of this region (Figure 8.1). In this case, the particle is trapped within the $V = 0$ "well", and the walls are infinitely high, thus there can be no tunneling. The problem reduces to solving the TISE within the well:

a.

b.

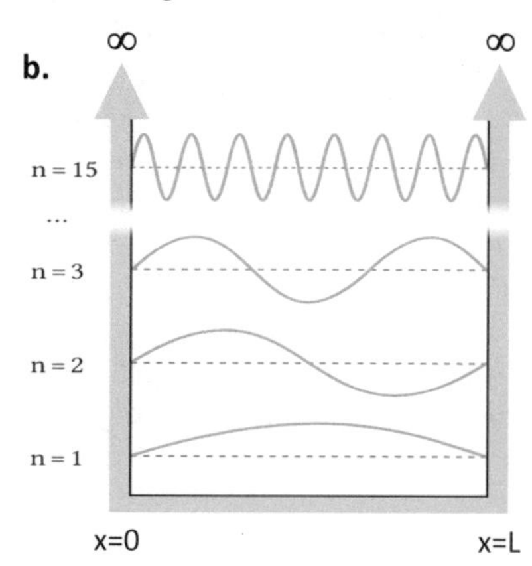

FIGURE 8.1 Infinite Square Well. **(a)** If a well is deep enough, you will not be getting out. **(b)** One-dimensionl infinite square well. The particle can move only along x. The infinite height of the well at $x < 0$ and $x > L$ prevents escape of the particle. The first three eigenfunctions ($n = 1, 2, 3$ using equation 8.3) and the n=15 eigenfunction are shown. Higher energy states have more humps (and more nodes) and shorter wavelength. [LEFT image by © cybrain, www.shutterstock.com/image-illustration/top-view-old-flooded-elevator-shaft-625771715. RIGHT image by © MikeRun, https://commons.wikimedia.org/wiki/File:Particle-in-a-box-wavefunctions.svg, licensed under CC BY-SA 4.0.]

DOI: 10.1201/9781003356172-10

$$[-\hbar^2/2m(\partial/\partial x)^2]\psi = E\psi \qquad (0 < x < L) \qquad (8.1)$$

This resembles the free particle problem, for which the solutions were either a complex exponential or sin or cos functions. However, ψ must be zero outside of the well where $V \rightarrow \infty$. In order for ψ to be continuous everywhere, ψ inside the well must go to zero at the boundaries ($x = 0$ and $x = L$). This condition at $x = 0$ eliminates the exponential and cos forms, therefore:

$$\psi = \sin(kx) \quad \{\text{inside the well only}\} \qquad (8.2)$$

Just as for the free particle, this solution will satisfy the TISE so long as $\hbar^2k^2/2m = E$. The sole remaining task is to satisfy the boundary condition at the right edge of the box:[1]

$$\psi(L) = 0 = \sin(kL) \qquad (8.3)$$

Since $\sin(\theta) = 0$ when $\theta = n\pi$, then:

$$kL = n\pi, \quad \text{or } k = n\pi/L \qquad (8.4)$$

This is our first case of quantization: k can only take on certain values, those values being separated by an amount π/L. Consequently $p = \hbar k$ and $E = p^2/2m$ can take on only certain distinct (quantized) values. This occurs because the particle is trapped in a well. It needn't have been an infinite well: the energies of a particle trapped in a finite well are also quantized. But the math is far simpler for an infinite well.

Note that the allowed energies ($E = h^2k^2/2m$) are proportional to n^2, where $n = 1,2,3$, etc. Note also that these solutions are all sin waves, with the number of "humps" equal to n (Figure 8.1b shows first three eigenstates). The number of nodes, or points where $\psi = 0$, is $n - 1$ (the null points at the boundaries are not counted as nodes). The number n is the first quantum number that we encounter. (See problem 8.2 to derive E as a function of n.)

In a hydrogen atom, the electron is trapped by attraction to the nucleus. In Chapter 16, we will show that only certain electronic energies are allowed for the hydrogen atom, that these energies are related to n^2, and that the number of nodes in the electronic orbitals is $n - 1$.

8.2 EIGENFUNCTION ↔ SEPARABLE ↔ STATIONARY STATE ↔ PURE STATE

We will normally not multiply $\psi(x)$ by $e^{-i\omega t}$ to form $\Psi(x,t)$. We will do so in this section merely to discuss the time-dependence of the probability P for an eigenfunction. For the PIB:

$$\Psi(x,t) = \sin(kx)\, e^{-i\omega t} \qquad (8.5)$$

$$P = |\Psi|^2 = \Psi^*\Psi = \sin^2(kx)\, e^{+i\omega t}\, e^{-i\omega t} = \sin^2(kx) \qquad (8.6)$$

While the wave function Ψ has time-dependence, the probability P does not! The probability distribution, describing the position of the particle, does not change with time. For this reason, eigenfunctions are also called "stationary states". This does not mean that the electron has no velocity: it clearly has momentum proportional to k. But its position, or rather its probability of being found at a particular position, does not change *with time*. This absence of time-dependence is a direct result of the separability of the eigenstate: the separable form allows the time part to cancel in equation 8.6. Therefore, for all practical purposes, the following six terms are synonymous: *eigenfunction, eigenstate, eigenvector, stationary state, separable solution*, and *pure state*.

Conversely, a state with a time-dependent probability must *not* be an eigenstate. It can also be called a *non-stationary state*, a *non-separable state*, or more commonly, a *superposition of states*, since it can be thought of as being comprised of a mixture of eigenstates.

FIGURE 8.2 Linear Combination. Equal parts golden retriever and poodle produce a goldendoodle, which ideally can retrieve objects without shedding. If dogs were quantum particles, then retrievers and poodles would be eigenfunctions (pure states), and the goldendoodle would be a superposition of states. However, an attempt to determine the breed of the goldendoodle would cause it to collapse into either a retriever state or a poodle state, with equal probability.

[LEFT photo by © Scott Beckner, www.flickr.com/photos/29949904@N00/1499716622.

CENTER photo by © Dawn Nuczek, www.flickr.com/photos/31064702@N05/402 3771641/.

RIGHT photo by © Gullpavon, https://en.wikipedia.org/wiki/Goldendoodle#/media/ File:Golden_Doodle_Standing_(HD).jpg, licensed under CC BY-SA 4.0.]

8.3 NORMALIZATION

To normalize the PIB eigenstates, the following integral must be evaluated:

$$\langle\Psi|\Psi\rangle = \int_{-\infty}^{+\infty} dx\ \psi^*(x)\psi(x) = \int_0^L dx\ \psi^*(x)\psi(x) = \int_0^L dx\ \sin^2(kx) \qquad (8.7)$$

where the change of the integration limits reflects the fact that ψ is zero outside of the well. To evaluate this integral, use the following trick: since $\sin^2(\theta) + \cos^2(\theta) = 1$, then over an integral number of humps in a sine function, the average value of $\sin^2(\theta) = \frac{1}{2}$. Therefore:

$$\int_0^L dx \sin^2(kx) = \int_0^L dx\,(1/2) = (1/2)\int_0^L dx = L/2 \quad (8.8)$$

To be normalized, this integral should = 1. Multiplication of ψ by $\sqrt{2/L}$ will produce the desired normalized eigenstate:

$$\psi = \sqrt{2/L}\,\sin(kx) \quad \text{[normalized eigenstate]} \quad (8.9)$$

8.4 ORTHOGONALITY

It turns out that the eigenstates for a given Hamiltonian are orthogonal to each other[2]. This simply means that the integral of their product is zero:

$$\langle\psi_i|\psi_j\rangle = 0 \qquad [\text{if } i \neq j] \quad (8.10)$$

This can be shown to be true for the case of the PIB eigenstates, using symmetry arguments or via evaluating the integrals (see problem 8.4).

8.5 ORTHONORMALITY

Since the eigenstates are orthogonal, then once normalized, they form an orthonormal set (orthogonal to each other, and each one normalized). This can be concisely symbolized as:

$$\langle\psi_i|\psi_j\rangle = \delta_{ij} \quad (8.11)$$

where δ_{ij} is the Kronecker delta, and takes the value 1 if $i = j$, and zero otherwise.

8.6 EXPECTATION VALUES

The expectation value of a quantity is the mean value one would expect for that quantity, given a particular wave function. For instance, $\langle x\rangle$ is the symbol for expectation value of x, which simply means the average position of the particle.

For state ψ_1 (Figure 8.1b, bottom trace), the average position of the electron is obviously at the center of the well: $\langle x\rangle = L/2$. In other cases, the average position may not be as obvious, but it can be calculated via the formula:

$$\langle q\rangle = \langle\Psi|\hat{Q}|\Psi\rangle \quad (8.12)$$

where q represents the quantity, and $\hat{Q}$ is the operator associated with the quantity.[3,4] In equation 8.12, the Dirac bra-ket notation has been expanded to include the operator between the bra and the ket. Explicitly, this notation means:

$$<\Psi|\hat{Q}|\Psi> = \int_{-\infty}^{+\infty} dx\, \Psi^* \hat{Q}\, \Psi \quad (8.13)$$

The operator is placed between the two eigenfunctions. It operates forward onto Ψ. The result is then multiplied by Ψ*, and then the integral is performed. We have introduced operators for energy and momentum (equation 4.1). Here, we introduce the operator for position $\hat{x}$: it is simply the variable x itself. So to find <x>, we perform:

$$<x> = <\Psi|\, \hat{x}\, |\Psi> = <\Psi|\, x\, |\Psi> \quad (8.14)$$

As another example, to find the expectation value of momentum, simply replace the $\hat{x}$ operator with the $\hat{p}$ operator:

$$<p> = <\Psi|\, \hbar/i(\partial/\partial x)|\, \Psi> \quad (8.15)$$

PROBLEMS

8.1 Use equations 7.17 and 7.18 to prove that there is zero probability to tunnel into an infinitely high barrier.

8.2 For an infinite square well: **(a)** Inside the well, what is the difference between KE and E? **(b)** Given your answer to part (a), write an equation relating E to momentum p. **(c)** Use equation 8.4 to find the allowed values of E as a function of n.

8.3 Why is there no n = 0 solution for the PIB? Hint: what would be the shape and size?

8.4 Use symmetry arguments (visual inspection) to show that $<\psi_1|\psi_2> = 0$ for a PIB.

8.5 For PIB eigenfunctions, mathematical calculation of <E> is simpler than calculation of <x> or <p>. Why is that the case? (the answer is mathematical)

8.6 Calculate the energy difference between the PIB first excited and third excited states.

8.7 Calculate <p> for: **(a)** the free particle wave function e^{ikx}; **(b)** the PIB ground state; **(c)** the PIB fifth excited state.

8.8 **(a)** Assume that a racquetball is a trapped particle in a one-dimensional well, and the width of the well is the width of a racquetball court. Determine the difference in velocity of the ground and first excited state of the racquetball. **(b)** Repeat this calculation for an electron trapped in a carbon atom, using the atomic diameter as the width of the well. **(c)** What do your answers to parts (a) and (b) tell you about the need to use QM for large vs. small objects?

NOTES

1. ψ' need not be continuous at the boundaries, since V is infinitely discontinuous at $x = 0$ and $x = L$.
2. Provided that they correspond to distinct eigenenergies: we needn't worry about this subtlety here.
3. The time parts would cancel in all expectation values we calculate, and therefore $\psi(x)$ may be used instead of $\Psi(x,t)$ to calculate expectation values.
4. If ψ is not normalized, then the expectation value calculation in equations 8.12-8.15 will be incorrect. It can be corrected by dividing the result by $\langle\Psi|\Psi\rangle$: $\langle q\rangle = \langle\Psi|\hat{Q}|\Psi\rangle / \langle\Psi|\Psi\rangle$

9 The Formal Postulates of Quantum Mechanics

Investigation of the PIB in Chapter 8 has brought to light quantum mechanical principles that apply also to other systems (other potentials). We now gather and list those principles. Much of classical mechanics can be summed up in Newton's three simple laws. There is no analogous short consensus list of the important laws of quantum mechanics. However, most of the general QM principles we need *can* be summarized in one page or so.

I. The TDSE describes the behavior of a particle with respect to space and time
   - $\mathcal{H}\Psi = i\hbar(\partial/\partial t)\Psi$

II. The TDSE solution $\Psi(x,t)$ describes all that can be known about the particle

III. General properties of Ψ
   - Ψ and Ψ' are single-valued, finite, continuous {exception: Ψ' discontinuous at ∞ discont. of V}
   - normalize: scale so that $\int dx|\Psi|^2 = 1$
   - probability: $P = |\Psi|^2$ {Ψ must be normalized}
   - Dirac bra-ket notation: $\int dx|\Psi|^2 = \langle\Psi|\Psi\rangle$ {bra times ket, integrated over all space}

IV. If $V = V(x)$, separable solutions $\Psi(x,t) = \psi(x)f(t)$ can be found {in principle}
   - TISE: $\mathcal{H}\psi_n(x) = E_n\psi_n(x)$; solutions depend on form of V {$E_n = \hbar\omega_n$}
   - SISE: $i\hbar(\partial/\partial t)f_n(t) = E_n f_n(t)$; all solutions of form $f_n(t) = e^{-i\omega_n t}$ {$E_n = \hbar\omega_n$}
   - solutions $\psi_n(x)$ are eigenstates, eigenfunctions, eigenvectors, stationary states, pure states
   - E_n are called Energy eigenvalues or eigenenergies
   - a complete set of orthonormal efs can be found {in principle}
     - *Orthogonal*: $\langle\psi_i|\psi_j\rangle = 0$ {for $i \neq j$}
     - *Orthonormal*: $\langle\psi_i|\psi_j\rangle = \delta_{ij}$ {orthogonal and normalized}

V. Superposition principle: any linear combination of efs ψ_n is also a valid solution of the TDSE
   - the general form for Ψ is thus $\Psi(x,t) = \Sigma_n c_n \Psi_n = \Sigma_n c_n \psi_n e^{-i\omega_n t}$

DOI: 10.1201/9781003356172-11

- the general Ψ is *not* separable: only pure efs are separable and solve the TISE
- the general Ψ, being a mixture of energy eigenstates, does not have a well-defined energy

VI. Each measurable variable q is associated with an operator $\hat{Q}$

- the only possible result of a single measurement of q is an eigenvalue of $\hat{Q}$

VII. The expectation value of a measurement need not equal an eigenvalue

- expectation value <q>: mean value of q expected from a series of measurements
- $\langle q \rangle = \langle \Psi | \hat{Q} | \Psi \rangle$ {if not normalized: $\langle q \rangle = \langle \Psi | \hat{Q} | \Psi \rangle / \langle \Psi || \Psi \rangle$}
- <q>: weighted sum: sum of eigenvalues times probability to find that eigenvalue
- $\langle q \rangle = \Sigma_n q_n |c_n|^2$ {if not normalized: $\langle q \rangle = \Sigma_n q_n |c_n|^2 / \Sigma_n |c_n|^2$}

Postulates I–IV have been covered in previous chapters. Postulates V, VI and VII will now be demonstrated by means of an example. This example will also demonstrate that a superposition of states is non-separable, and that time parts of Ψ need not be included in expectation value calculations.

9.1 EXAMPLE: PIB SUPERPOSITION OF STATES

An eigenfunction of $\mathcal{H}$ is called a pure state, denoting that it is composed of only one eigenfunction (ef). A new state can be constructed by linearly combining efs. For instance, combining three parts n = 1 and two parts n = 5 PIB states produces:

$$\psi = 3\psi_1 + 2\psi_5 \tag{9.1}$$

First, let's normalize this state. Assume that ψ_1 and ψ_5 are normalized. Then clearly ψ is too large by some factor related to the coefficients 3 and 2. To normalize, evaluate $\langle \psi | \psi \rangle$:

$$\langle \psi | \psi \rangle = \{3\langle \psi_1 | + 2\langle \Psi_5 |\}\{3|\psi_1\rangle + 2|\psi_5\rangle\} =$$
$$9\langle \psi_1 | \psi_1 \rangle + 4\langle \psi_5 | \psi_5 \rangle + 6\langle \psi_1 | \psi_5 \rangle + 6\langle \psi_5 | \psi_1 \rangle$$

Use of the orthonormality condition (equation 8.10) reduces this to $\langle \psi | \psi \rangle = 13$. The required normalization constant is therefore $1/\sqrt{13}$.

ψ in equation 9.1 is called a superposition of states since it is mixture of efs. By definition, then, it is not a solution to the TISE, which can only be solved by individual efs. ψ *is* a solution to the TDSE, *almost*. Actually, the state in equation 9.1 cannot exist because, being a superposition of states, it is not separable. Each part, ψ_1 and ψ_5, should be multiplied by $e^{-i\omega t}$, but with distinct values of ω since the energies differ for ψ_1 and ψ_5. Thus, the normalized state should be written:

$$\Psi = (1/\sqrt{13})\{3\psi_1 e^{-i\omega_1 t} + 2\psi_5 e^{-i\omega_5 t}\} \tag{9.2}$$

Clearly, Ψ is not separable, since it has two distinct space parts multiplied by two distinct time parts. Therefore, Ψ cannot even be tested as a solution for the TISE. However, Ψ *does* solve the TDSE, as can be easily verified by plugging it into equation 5.1 (see problem 9.2). This is the crux of postulate IV above: a linear combination of efs is a solution to the TDSE. In fact, we state without proof that *all* possible solutions to the TDSE can be expressed as a linear combination of the efs.

What energy is associated with Ψ? Since it contains a mixture of ψ_1 and ψ_5, it must have some energy character of each. Since the coefficient of ψ_1 is larger, the energy should be closer to E_1 than to E_5. However, postulate V states that a single measurement of E can only return an E eigenvalue. In this case, a single measurement can only return either E_1 or E_5. If we were to create a large number of such states, and measure E for each of them, we would find the value E_1 more often than E_5, by a ratio of 9:4, which is just 3^2:2^2, or the square of the ratio of the coefficients in equation 9.2. Noting that probability is related to Ψ^2, this indicates that the coefficients, through a probability calculation, provide the likelihood of finding the particle in ψ_1 versus ψ_5. Averaging the energy measurements yields:

$$E_{ave} = (9E_1 + 4E_5)/13 \tag{9.3}$$

This weighted sum indicates that the average energy is closer to E_1 than to E_5, as required. There is a name for this type of averaging in QM: *expectation value*.

Postulate VI provides a simpler procedure to calculate the expectation value, which in this case would require evaluation of the following integral:

$$\begin{aligned} <E> = <\Psi|\mathcal{H}|\Psi> = \left(1/\sqrt{13}\right)\left\{3 <\psi_1|e^{+i\omega_1 t} + 2 <\psi_5|e^{+i\omega_5 t}\right\} \\ \mathcal{H}\left(1/\sqrt{13}\right)\left\{3|\psi_1> e^{+i\omega_1 t} + 2|\psi_5> e^{+i\omega_5 t}\right\} \end{aligned} \tag{9.4}$$

where the time functions have been removed from the bra-kets, since the integration is over space. Upon removal from the bra, each time part must be made into its complex conjugate. This is the reason for the positive exponentials. Next, the operation of $\mathcal{H}$ onto ψ_1 and ψ_5 can be replaced by $E_1|\psi_1>$ and $E_5|\psi_1>$ by use of the TISE:

$$<E> = (1/13)\ \{3<\psi_1|e^{+i\omega_1 t} + 2<\psi_5|e^{+i\omega_5 t}\}\ \{3E_1|\psi_1>e^{-i\omega_1 t} + 2E_5|\psi_5>e^{-i\omega_5 t}\} \tag{9.5}$$

Multiplication will result in four terms. The two cross terms disappear due to orthogonality. In the two other terms, the time parts cancel to give $e^0 = 1$. Thus, all time parts disappear, and, keeping only the non-cross terms, and using the normality condition in the final step, we have:

$$<E> = (1/13)\ \{3<\psi_1|3E_1|\psi_1> + 2<\psi_5|2E_5|\psi_5>\} = (1/13)\{9E_1 + 4E_5\} \tag{9.6}$$

which is identical to equation 9.3. This result demonstrates the usefulness of postulate VI, and also points out that the time parts can be ignored in expectation value calculations: ψ from equation 9.1 could have been used instead of Ψ. This greatly simplifies the calculation (see problem 9.3). The expectation value can also be more easily calculated using the last line of postulate VII on page 46 (see problem 9.5) .

9.2 TIME-DEPENDENCE OF A SUPERPOSITION OF STATES

While the time part of Ψ is not needed to calculate <E>, and hence <E> has no time dependence, this does not mean that the probability P has no time dependence. A superposition of states is by definition a non-stationary state, which means that its probability *must* have time dependence. In this case:

$$\begin{aligned} P(x,t) = |\Psi^2| &= (1/13)\ \{3\psi_1^* e^{+i\omega_1 t} + 2\psi_5^* e^{+i\omega_5 t}\}\ \{3\psi_1 e^{-i\omega_1 t} + 2\psi_5 e^{-i\omega_5 t}\} \\ &= (1/13)\{9\psi_1^*\psi_1 + 4\psi_5^*\psi_5 + 6\psi_1^*\psi_5\, e^{+i(\omega_1-\omega_5)t} \\ &\quad + 6\psi_5^*\psi_1\, e^{+i(\omega_5-\omega_1)t}\} \end{aligned} \tag{9.7}$$

Focusing on the time-dependence, we note that the cross terms ($\psi_1^*\psi_5$ and $\psi_5^*\psi_1$ terms) have complex exponentials which oscillate at a frequency equal to the difference in frequencies of the component waves. This phenomenon has analogies in classical mechanics, where it is called interference, or beats. Here, it is our first observation of *quantum interference*: the two component waves, at frequency ω_1 and ω_5, interfere to produce probability beats. Thus the particle's positional probability changes with time, although the average energy does not.

9.3 HILBERT SPACE

David Hilbert was a German mathematician who introduced the concept of an infinite dimensional space to help visualize the use of a basis set. In QM, for instance in the case of the PIB, the infinite set of all efs $\psi_1, \psi_2, \psi_3 \ldots$ is called the PIB Hilbert space. Any solution to the TDPT can be described as a projection into the PIB Hilbert space. For instance, the solution in equation 9.2 is a projection of $(3/\sqrt{13})$ along the ψ_1 direction, $(2/\sqrt{13})$ along the ψ_5 direction, and no projection along the remaining directions. Note that the total length of the equation 9.2 "vector" in Hilbert space, obtained by summing the square of the coefficients, is 1, since this wave function was normalized.

PROBLEMS

9.1 Construct a vector consisting of three meters along x and two meters along y. **(a)** What is the length of the vector? **(b)** By what factor would you need to decrease the length in order to make the vector of unit length (1m)? **(c)** Compare you answers with the discussion for the superposition of states in this chapter. Use the term *orthogonal* in your discussion.

9.2 Show that the wave function in equation 9.2 is *not* an eigenfunction of the TISE, but *is* an eigenfunction of the TDSE. In the TISE calculation only, assume t=0 for simplicity.

9.3 **(a)** Use the form of ψ (x) in equation 9.2 to show, using a simpler derivation than shown in the text, the value of <E>. **(b)** Leave off the time factors in equation 9.2 and redo the calculation from part (a). **(c)** From your results in parts (a) and (b), make a statement regarding the preferred method to calculate expectation values: should the time parts be included or not?

9.4 Perform an internet search to find a formula relating acoustic beats to the frequency of two notes played at the same time. Compare this result with equation 9.7.

9.5 Use the two equations in postulate VII on page 46, containing the coefficients c_n, to calculate <E> for the wave function in equations 9.1 and 9.2.

9.6 Calculate the expectation value of momentum for the following free particle wave function:

$$\Psi(x,0) = 4e^{ik1x} + 2e^{ik2x} + 7e^{ik3x}$$

For purposes of the problem, assume that each of the exponential functions (e^{ikx}) is normalized, and provide your answer in terms of the arbitrary numbers k_1, k_2 and k_3.

10 Simple Harmonic Oscillator (SHO): $V = \frac{1}{2} kx^2$

The one-dimensional simple harmonic oscillator (SHO) is a useful approximation for a chemical bond. A bond can be envisioned as a spring connecting two atoms: if the atoms move too far apart, or move too close together, they are pulled/pushed back toward the equilibrium position (the bond length). Treatment of the quantum SHO shows that molecular vibration energy levels are evenly spaced, and that bonds always vibrate, even at T = 0 K. Analysis of the SHO also introduces the creation and annihilation operators as a convenient way to obtain eigenstates and eigenenergies.

10.1 CLASSICAL SHO (MASS ON A SPRING, BALL IN A WELL)

For comparison with the results of the quantum SHO, we briefly treat the classical SHO, which is defined by the potential $V(x) = \frac{1}{2} kx^2$ (see figure 10.1). A mass on a spring provides an example of classical SHO motion. In classical mechanics, motion can be determined by using Newton's 2nd law, F = ma. Since the relation between force and potential is $F = -\partial V/\partial x$, we find that for the SHO potential, F = -kx. We also use the definition of acceleration *a* as the second derivative of position to recast Newton's 2nd law into differential form:

$$F = ma \rightarrow -kx = m\, dx^2/dt^2 \tag{10.1}$$

where *x* is the position of the mass as a function of time. The task is to find a function *x(t)* that solves the above differential form of Newton's 2nd law. The function must reproduce itself upon being differentiated twice, apart from some constants. x = Acos(ωt) works. Plugging in produces:

$$-kAcos(\omega t) = mA\left(-\omega^2\right)cos(\omega t) \rightarrow x = -A\cos(\omega t) \text{ satisfies the equation if } \omega = \sqrt{k/m} \tag{10.2}$$

This solution indicates that the mass oscillates back and forth with a frequency ω that depends only on the force constant of the spring (*k*) and the mass of the object (*m*). The frequency is the same regardless of the amplitude A. This is why a grandfather

DOI: 10.1201/9781003356172-12

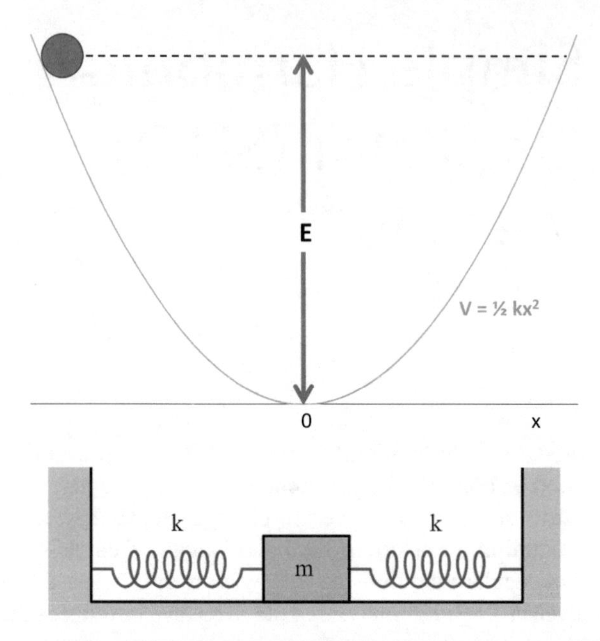

FIGURE 10.1 Classical one-dimensional SHO. A particle free to move only along x is constrained by a potential $V(x) = \frac{1}{2} kx^2$. **(a)** The shape of the potential is graphed as the curved line. Although the particle only moves in one dimension, we can understand the motion by referring to a ball rolling down a hill of the same shape as V(x). The ball will move fastest at the bottom of the hill, and will rise to the same height on the other side of the hill. **(b)** A mass connected to two walls by two springs would feel an SHO potential, and would oscillate back and forth about its equilibrium position at x = 0, with the same adjustments in velocity as discussed for the ball in part (a). [Part (b) image by © MikeRun, https://commons.wikimedia.org/wiki/File:Horizontal-mass-on-two-springs.svg#/media/File:Horizontal-mass-on-two-springs.svg, licensed under CC BY-SA 4.0.]

clock keeps time so well: the pendulum follows SHO mathematics, and thus the oscillation takes a constant time, even as the amplitude reduces with time, due to friction at the fulcrum.

A ball on a hill analogy is useful (figure 10.1 a). If a ball is placed at a point in the potential and allowed to roll down, it would roll back and forth at a constant frequency, regardless of the height at which it was released. Furthermore, the ball would be least likely to be found at the very bottom of the well, and most likely to be found at its highest points, since it travels fastest at the bottom, and actually stops momentarily at the top of its motion. The total energy E can be calculated from the potential energy when the ball stops at its highest point: E = mgh. There is of course no quantization classically: any amplitude is permitted, including zero energy (ball resting at the lowest point).

Note that for this classical solution, we know exactly where the mass or the ball is at all times. The function *x(t)* is *not* a probability function; it is an *exact* description of the position and hence also of momentum at any time (see problem 10.1). Note also that k and ω here are *not* the QM k and ω that represent momentum and energy. Here, k is the force constant, and can have only one value for a given spring, while $\omega = \sqrt{k/m}$, and likewise has only a single value for a given spring/mass combination.

10.2 QUANTUM SHO

Inserting $V = ½\, kx^2$ into the TISE produces:

$$[-\hbar^2/2m(\partial/\partial x)^2 + ½\, kx^2]\psi = E\psi \tag{10.3}$$

The task is to find the eigenfunctions $\psi(x)$ that solve this equation. This is somewhat analogous to finding $x(t)$ in the classical case. However, $\psi(x)$ represents the *probability* to find the particle at position x, rather than the precise position as a function of time. The form of the TISE is simplified by the following variable substitutions, and isolation of the double derivative:

$$\zeta = x\sqrt{(m\omega_o/\hbar)},\ \varepsilon = E/\hbar\omega_o \tag{10.4}$$

$$\rightarrow d^2\psi/d\zeta^2 = (\zeta^2 - 2\varepsilon)\psi \tag{10.5}$$

where again, k and ω_o are *not* the k and ω in previous chapters: here they represent the force constant and the frequency of the SHO, respectively. In this simplified TISE of equation 10.5, also called the dimensionless TISE for an SHO, just think of ζ as being a different form of x, almost like a conversion from feet to meters: it still represents position. Likewise, ε still represents an energy. The TISE in this form still states that KE + V = E (can you spot which term is which?).

The SHO TISE (equation 10.5) is not straightforward to solve. However, it *is* simple to solve when $\zeta >> \varepsilon$; that is, far from the center of the well, or for very low energy. Under these conditions, the TISE reduces to:

$$d^2\psi/d\zeta^2 = (\zeta^2)\psi \tag{10.6}$$

which is solved by:

$$\psi \sim \exp(-\zeta^2/2) \quad \text{as} \quad \zeta \rightarrow \infty \tag{10.7}$$

Proof that this solves equation 10.6 is shown in problem 10.2. Therefore, this solution should work for lowest ε, and in fact, it is the ground state solution ψ_0. Plugging ψ_0 into the full SHO TISE (equation 10.5) shows that it is a solution so long as $\varepsilon = ½$. Therefore, the ground state energy is $E = ½\hbar\omega_o$ (see equation 10.4).

Next, note that even for higher energies, this exponential solution should be approximately correct at large ζ. Therefore, we surmise that each eigenfunction should contain this same exponential squared factor. But at small ζ, the 2ε term in equation 10.5 becomes important. Therefore, the eigenfunctions must also have a

TABLE 10.1
Hermite polynomials

$H_0(x) = 1$
$H_1(x) = 2x$
$H_2(x) = 4x^2 - 2$
$H_3(x) = 8x^3 - 12x$
$H_4(x) = 16x^4 - 48x^2 + 12$
$H_5(x) = 32x^5 - 160x^3 + 120x$
$H_6(x) = 64x^6 - 480x^4 + 720x^2 - 120$

TABLE 10.2
QM SHO wave functions (not normalized)

n	E_n	ψ_n
0	$\hbar\omega_0/2$	$e^{-\zeta^2/2}[\,1\,]$
1	$3\hbar\omega_0/2$	$e^{-\zeta^2/2}[\,2\zeta\,]$
2	$5\hbar\omega_0/2$	$e^{-\zeta^2/2}[\,4\zeta^2 - 2\,]$
3	$7\hbar\omega_0/2$	$e^{-\zeta^2/2}[\,8\zeta^3 - 12\zeta\,]$
4	$9\hbar\omega_0/2$	$e^{-\zeta^2/2}[\,16\zeta^4 - 48\zeta^2 + 12\,]$
5	$11\hbar\omega_0/2$	$e^{-\zeta^2/2}[\,32\zeta^5 - 160\zeta^3 + 120\zeta\,]$
6	$13\hbar\omega_0/2$	$e^{-\zeta^2/2}[\,64\zeta^6 - 480\zeta^4 + 720\zeta^2 - 120\,]$

second part that becomes important at small ζ. We assume this second part is a polynomial $P(\zeta)$, and then we seek the form of the polynomial:

$$\psi = \exp(-\zeta^2/2)\,P(\zeta) \tag{10.8}$$

Fortuitously, a French mathematician named Hermite solved a related differential equation in the mid-19th century. His solutions, called the Hermite polynomials (table 10.1), when multiplied by the squared exponential term, are the SHO efs we seek (table 10.2). Plugging these solutions into equation 10.5 reveals that the nth order solution $\{\exp(-\zeta^2/2)\,P_n(\zeta)\}$ has energy eigenvalue $\varepsilon = n + ½$, or (via equation 10.4):

$$E_n = (n + ½)\,\hbar\omega_o \tag{10.9}$$

Thus, the energies of the quantum SHO efs increase linearly with n (different from square well): the levels are evenly spaced. Also interestingly, there is no state corresponding to E = 0. The ground state energy is ½ $\hbar\omega_o$. So even at a temperature of absolute zero, a chemical bond, which approximates an SHO, must have

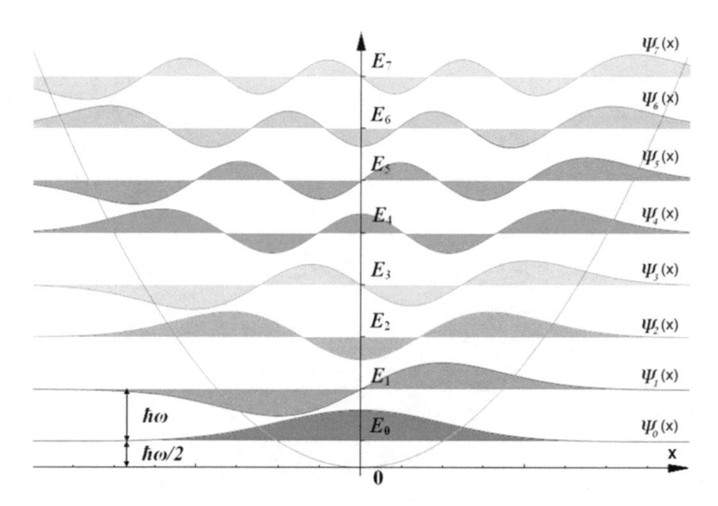

FIGURE 10.2 **Quantum SHO**. The potential V = ½ kx^2 leads to a series of eigenfunctions with an increasing number of humps. The energy levels are evenly spaced. [Image by © AllenMcC, https://en.wikipedia.org/wiki/File:HarmOsziFunktionen.png, licensed under CC BY-SA 3.0]

some vibrational energy. This minimal amount of energy (½ $\hbar\omega_0$), called the *zero point energy*, is clearly a *quantum* phenomenon: a *classical* pendulum *can* have zero amplitude.

The wave functions are shown graphically in figure 10.2. As expected, they decay exponentially far from the center, but resemble an nth order polynomial at small ζ. Each solution ψ_n contains an nth order Hermite polynomial, and therefore it contains n nodes. The similarity and differences from the PIB efs are discussed in problem 10.6.

10.3 CREATION/ANNIHILATION OPERATORS

There is a far simpler procedure to generate the quantum SHO efs. The only justification we provide for this procedure is that it works.

First assume as we did above that the ground state is:

$$\psi_0 = \exp(-\zeta^2/2) \tag{10.10}$$

Substitution into the TISE (equation 10.5) shows that this is a valid solution with eigenvalue $\varepsilon = ½$. Now define two new operators a^+ and a^-:

$$a^+ = [d/d\zeta - \zeta]; \quad a^- = [d/d\zeta + \zeta] \tag{10.11}$$

and note that $a^+\psi_1 = \psi_2$, and similarly that $a^-\psi_2 = \psi_1$. (see Problem 10.5). In fact, it can be shown that in general (apart from normalization):

$$a^+\psi_n = \psi_{n+1}; \quad a^-\psi_n = \psi_{n-1} \tag{10.12}$$

FIGURE 10.3 Creation and annihilation.

[TOP photo from https://en.wikipedia.org/wiki/The_Creation_of_Adam#/media/File:Michelangelo_-_Creation_of_Adam_(cropped).jpg, public domain. BOTTOM photo by © JasonWeingart, https://commons.wikimedia.org/wiki/File:Evolution_of_a_Tornado.jpg, licensed under CC BY-SA 4.0.]

The two operators, a^+ and a^-, are called the raising and lowering operators, because operation on ψ raises or lowers the quantum number n and hence also the energy. These operators are convenient tools to generate the efs. You need only have the form of one of the ψ_n, and the other efs can be generated by serial application of a^+ or a^-. These operators are also called *creation* and *annihilation* operators, since they "create" or "destroy" energy. We introduce these two operators here only as a curiosity, and for an opportunity to practice using operators. Also, they may be useful on an exam: you needn't memorize the SHO eigenfunctions: you can derive them!

PROBLEMS

10.1 For a classical SHO with solution $x = Acos(\omega t)$ and amplitude 0.6 meters: **(a)** Calculate the maximum value of momentum p. **(b)** At what position (what value of x) is this maximum momentum achieved? **(c)** At what position is the momentum zero?

10.2 Prove that the SHO wavefunction given in equation 10.7 solves the simplified TISE in the form given in equation 10.6 when $\zeta \rightarrow \infty$.

10.3 **(a)** Prove that the SHO wavefunction given in equation 10.7 solves the TISE in the form given in equation 10.5, but only if $\varepsilon = \frac{1}{2}$. **(b)** Use equation 10.4 to find the corresponding value of E.

10.4 **(a)** Prove that the n = 1 SHO solution from table 10.2 solves the TISE in equation 10.5. **(b)** Determine the value of ε. **(c)** Determine the value of E.

10.5 A pendulum is an example of SHO motion. Thus, according to QM theory, a pendulum *cannot* remain still. **(a)** Use equation 10.9 to calculate the zero point energy (ground state SHO energy) of a pendulum. **(b)** Calculate the maximum momentum p of the pendulum in the SHO ground state. **(c)** Calculate the maximum amplitude in the SHO ground state. For this problem, assume that the mass at the end of the pendulum is 2 grams, and that the force constant is 500 Newtons/meter.

10.6 **(a)** Use the creation operator to create the SHO n = 3 state from the n = 2 state. **(b)** Use the annihilation operator to derive the SHO n = 2 state from the n = 3 state. **(c)** Describe why the mathematical forms of the creation and annihilation operators would produce new functions with the required polynomial degree.

10.7 Discuss the similarities and differences between the appearance of the SHO and PIB efs.

10.8 The "*correspondence principle*" states that, at high n, quantum and classical descriptions become similar if not indistinguishable. For the SHO probability distribution: **(a)** show that low energy QM solutions differ from the classical solution; **(b)** show that high energy QM solutions look more like the classical solution.

Approximation Methods

11 Time-Independent Perturbation Theory (TIPT)

11.1 THREE APPROXIMATION METHODS

We have now solved the TISE for a free particle, a particle encountering rectangular barriers including an infinite well (PIB), and a particle in an SHO potential. These are the only one-dimensional potentials that we will treat exactly. Most potentials give rise to a TISE for which an exact solutions cannot be found. This does not mean that we are powerless. If the Hamiltonian *resembles* a Hamiltonian for which we do have solutions, then the difference can be treated as a small *perturbation*.

If the small perturbation is constant in time, then its effect will be to slightly alter the eigenenergies and eigenfunctions. This is the topic of *time-independent perturbation theory (TIPT)*. If, instead, the perturbation is turned on and then off, the system in the end has the same eigenstates and eigenenergies as it did before the perturbing event. The question in that case becomes one of transitions: if the particle began in state 3, for instance, what is the probability that afterwards, due to the temporary perturbation, it has jumped to state 4? This is the topic of *time-dependent perturbation theory (TDPT)* (see Chapter 12). Finally, even if the Hamiltonian does not resemble a well-studied Hamiltonian, it may still be possible to approximate the eigenenergies and eigenstates, via the *Variational Method* (see Chapter 13).

These approximation methods are central to quantum chemistry: for instance, the effects of multiple atomic electrons on each other, or of a magnetic field on a molecule, are treated via these approaches.

11.2 TIPT (TIME-INDEPENDENT PERTURBATION THEORY)

Suppose that an electron is subject to a potential that we know how to treat, like for instance a particle in a box potential, but that the box is slightly altered. For instance, perhaps the floor is slightly raised above V = 0, or it is slightly slanted (V = 0 on the left edge, but V rises slightly as you move to the right). It is a good assumption that the eigenstates and eigenenergies will be similar to those for a PIB, but with some small change. How can the change be approximated?

The derivation is tricky, but the results are simple. We merely state the results, and give an example of how to use them.

DOI: 10.1201/9781003356172-14

FIGURE 11.1 The Princess and the Pea. A small perturbation affects the state of the Princess. [Image by © KateGabriel, https://kategabrielle.com/products/the-princess-and-the-pea-art-print, by permission.]

First lets set up the problem:

- assume an electron is subject to a perturbed Hamiltonian: $\mathcal{H} = \mathcal{H}_o + \mathcal{H}_1$
- $\mathcal{H}_o$ is a standard potential (e.g., PIB)
- $\mathcal{H}_1$ is a small perturbation (e.g. slanted floor)
- let $\psi_{oi}(x)$ be the known efs of $\mathcal{H}_o$ with e.v. E_{oi} (o = original, *not* g.s.)
- starting from these, we can approximate the new efs ϕ_i and e.v.s of the perturbed $\mathcal{H}$

Approximate change in Eigenenergies:

$$\Delta E_i = \langle \psi_{oi} \mid \mathcal{H}_1 \mid \psi_{oi} \rangle \equiv \mathcal{H}_{1ii} \tag{11.1}$$

Approximate change in Eigenstates:

$$\phi_i = \psi_{oi} + \Sigma_{k \neq i} \left[\mathcal{H}_{1ki} / (E_{oi} - E_{ok}) \right] \psi_{ok} \tag{11.2}$$

Equation 11.1 simply states that the eigenenergies change by an amount equal to the expectation value of the perturbation itself, where the original eigenfunctions are used in the calculation. For instance, the energy of the 3rd ef changes by an amount $\mathcal{H}_{133}$. Equation 11.2 states that the new eigenstates (ϕ_i) look like the old eigenstates (ψ_{oi}), but there is a small amount of the other original eigenstates "mixed in". For instance, the new 3rd ef is old 3rd ef plus a bit of each old efs number 1, 2, 4, 5, 6, 7 … .

How much of each of these old efs mix in to form the new 3rd ef? This depends on two factors: $\mathcal{H}_{1ki}$ and $(\mathbf{E}_{oi} - \mathbf{E}_{ok})$. For instance, to calculate how much old 4th ef mixes in to help form the new 3rd ef, we must calculate:

$$\mathcal{H}_{143} = \langle\psi_{o4}|\, \mathcal{H}_1 |\psi_{o3}\rangle \quad \text{and} \quad (\mathbf{E}_{o3} - \mathbf{E}_{o4}) \tag{11.3}$$

If $\mathcal{H}_{143}$ does not equal zero, we say that the perturbation $\mathcal{H}_1$ "connects" states 3 and 4. It causes them to "mix": the new state 3 has some old state 4, and the new state 4 will also have some old state 3. The presence of the energy difference in the denominator in equation 11.2 defines a tendency for states of similar energy to mix more strongly than those with very different energies.

11.3 MATRIX SHORTHAND

Changes in *eigenenergies* thus are found by calculating integrals such as $\mathcal{H}_{133}$ (with matching indices) while changes in *eigenstates* are found by calculating integrals such as $\mathcal{H}_{143}$ (with non-matching indices). To help remember this, it is convenient to think of a series of $\mathcal{H}_1$ integrals forming a matrix:

$$\begin{pmatrix} \mathcal{H}_{111} & \mathcal{H}_{112} & \mathcal{H}_{113} \\ \mathcal{H}_{121} & \mathcal{H}_{122} & \mathcal{H}_{123} \\ \mathcal{H}_{131} & \mathcal{H}_{132} & \mathcal{H}_{133} \end{pmatrix} \tag{11.4}$$

The diagonal elements, which are expectation values, give the energy changes. The off-diagonal elements each specify whether the perturbation "connects" two states. For instance, if the matrix is calculated as the following (assume units of electron volts, or eV):

$$\begin{pmatrix} 0.2 & 0.0 & 0.0 & 0.0 \\ 0.0 & 0.1 & 0.0 & 0.3 \\ 0.0 & 0.0 & -0.2 & 0.0 \\ 0.0 & 0.3 & 0.0 & 0.1 \end{pmatrix} \tag{11.5}$$

then the perturbation increases the 2nd energy by 0.1 eV, and no states mix, except for states 2 and 4.

11.4 TIPT EXAMPLE: PIB PERTURBED BY AN ELECTRIC FIELD ε

PROBLEM

Suppose that a charged particle in a box (PIB) is perturbed by a small uniform electric field of strength ε. Calculate the change in the PIB eigenenergies due to the perturbation.

SOLUTION

First convert the electric field to a potential energy. Using classical electromagnetism, we have:

$$F = q\varepsilon = -(\partial V/\partial x) \rightarrow V_1 = -q\varepsilon x = \mathcal{H}_1 \tag{11.6}$$

If this is confusing, just assume that the perturbation is given as $\mathcal{H}_1 = -q\varepsilon x$. Then, using equation 11.1, we have:

$$\begin{aligned} \Delta E_i = \mathcal{H}_{1ii} &= \langle \psi_{oi} | -q\varepsilon x | \psi_{oi} \rangle = -q\varepsilon \langle \psi_{oi} | x | \psi_{oi} \rangle \\ &= -q\varepsilon \langle x \rangle = -q\varepsilon\,(L/2) \end{aligned} \tag{11.7}$$

where the constants q and ε were removed from the integral, and we recognize that the remaining integral is just an expectation value of x, which is $L/2$ (the mean position of the unperturbed particle is in the center of the box of width L). Thus, all eigenenergies (n = 1, 2, etc.) are perturbed by the same amount, and that amount is proportional to the charge on the particle, the electric field strength, and the width of the box.

It is left as an exercise to calculate examples of how eigenstates change due to a time-independent perturbation.

PROBLEMS

11.1 Set up integrals (use bra-ket notation if you like) that provide the following information, but do not do the integration: **(a)** the change in E of state n = 3 of an SHO due to a constant electric field perturbation; **(b)** the new state n = 3 that results from the same perturbation; **(c)** a test of whether the SHO states n = 2 and n = 5 mix due to the perturbation.

11.2 Suppose you perform all of the required integrals to evaluate the effect of a constant electric field perturbation on an SHO (see problem 11.1), and from these you determine that none of the energies are altered by the perturbation, except the n = 4 state which is reduced by 0.33 eV, and the n = 2 state which increases by the same amount. Also, you determine that none of the states mix, except for states n = 2 and n = 4. Draw a perturbation matrix (see equations 11.4 and 11.5) that incorporates all of this information.

11.3 Determine the perturbation matrix (see equations 11.4 and 11.5) for a PIB subject to a constant perturbation $V_0 = 12.6$ eV.

11.4 A PIB is subject to the time-*independent* perturbation: $\mathcal{H}_1 = \sin(2\pi x/L)$. **(a)** Does the ground state energy change? Explain. **(b)** Does the first excited state energy change? Explain. HINT: In each case, set up an integral and use visual symmetry arguments to evaluate the integral.

12 Time-Dependent Perturbation Theory (TDPT)

In Chapter 11, the case of a V(x) slightly different from a standard potential (PIB) was treated. The perturbation did not change with time: in the example, the electric field was always on. What if, instead, the electric field is *turned on* for a few seconds and is then *turned off?* This situation is treated via TDPT.

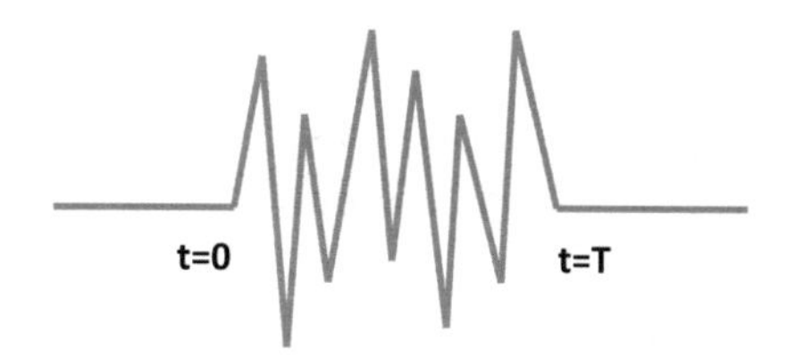

FIGURE 12.1 Time-dependent perturbation. The perturbation is present only between time t = 0 and t = T.

12.1 TDPT

First let's set up the problem:

- assume an electron is subject to a perturbed Hamiltonian $\mathcal{H} = \mathcal{H}_0 + \mathcal{H}_1(t)$
- $\mathcal{H}_0$ is a standard potential we can solve (such as PIB)
- $\mathcal{H}_1$ is small perturbation that is turned on (at time t = 0), then off (at time t = T)
- we need not know exactly what is happening at any time between t = 0 and t = T. We only wish to know the state of the system at t = T, due to the perturbation that has just ended
- specifically, if the system was, for instance, in state ψ_3 before the perturbation, what is the probability that, after the perturbation, it is now in state ψ_4?

Again, the derivation is complicated, but the result is (relatively) simple. We merely state the result and illustrate its use with two examples:

DOI: 10.1201/9781003356172-15

$$P(i \rightarrow j) = |a_{ij}|^2 \quad \text{(transition probability)} \tag{12.1}$$

$$a_{ij} = \delta_{ij} - i/\hbar \int_0^T dt \langle \psi_{oj} | \mathcal{H}_1(t) | \psi_{oj} \rangle e^{(i/\hbar)(E_{oj} - E_{oi})t} \tag{12.2}$$

or in simpler notation:

$$\mathbf{a_{ij} = \delta_{ij}} - i/\hbar \int_0^T \mathbf{dt}\, \mathcal{H}_{1ji}\, \mathbf{e}^{i(\omega_{oj} - \omega_{oi})t} \quad \text{(transition amplitude)} \tag{12.3}$$

- $P(i \rightarrow j)$: probability that a particle in state i transitions to state j due to $\mathcal{H}_1(t)$
- $P(i \rightarrow j)$: calculate by squaring $\mathbf{a_{ij}}$, analogous to squaring a wave function
- $\mathbf{a_{ij}}$ can be thought of as a transition amplitude
- the $\boldsymbol{\delta_{ij}}$ term suggests that a particle starting in ψ_3 will most likely end up in ψ_3
- the time integral sums the effect of the perturbation over time
- $\mathcal{H}_{1ji}$ is an off-diagonal matrix element as discussed in Chapter 11 (a space integral). It represents the ability of $\mathcal{H}_1$ to connect ψ_i and ψ_j
- The exponential term is a beat frequency: the effect depends on the value of $(\boldsymbol{\omega_{oj} - \omega_{oi}})$T. Simply put, this indicates that the probability for the particle to transition to other states sloshes back and forth if the perturbation remains on long enough. But, somewhat paradoxically, the sloshing is faster to more distant states (with a much different energy).

12.2 TDPT EXAMPLE 1: PIB TEMPORARILY PERTURBED BY AN ELECTRIC FIELD

PROBLEM

Suppose that a charged particle in a box is perturbed by a small uniform electric field of strength $\boldsymbol{\varepsilon}$. This electric field is turned on for a short time T, and then is turned off. Determine what sorts of transitions can be induced by this perturbation.

$$\mathcal{H}_1(t) = -q\varepsilon x \quad \text{(on from } t = 0 \text{ to } t = T) \tag{12.4}$$

SOLUTION

First do the space integral that is inside the time integral in equation 12.2:

$$\mathcal{H}_{1ji} = \langle \psi_{oj} | \mathcal{H}_1(t) | \psi_{oi} \rangle = \langle \psi_{oj} | -q\varepsilon x | \psi_{oi} \rangle = -q\varepsilon \langle \psi_{oj} | x | \psi_{oi} \rangle \tag{12.5}$$

The last term *almost* looks like an expectation value of x. But it is an *off-diagonal* element, while $\langle x \rangle$ is the *diagonal* element. Apparently, states that can be transitioned between must have a non-zero off-diagonal matrix element of x. Thus, we must solve the integral:

$$<\psi_{oj} | x | \psi_{oi}> = \int_o^L dx \sin(k_j x)\, x \sin(k_i x) \tag{12.6}$$

Solving this integral requires a series of trigonometry and calculus tricks. This math is unimportant to us, we simply state the result: the integral vanishes if *i-j* is even. This immediately yields the interesting result that *only odd transitions are allowed (those where the state changes by 1,3,5, ... steps).*

This is our first encounter with a *selection rule*: *only certain transitions are allowed.* Specifically, we now know that a weak, uniform electric field applied to a PIB for a short period of time *cannot* cause the n = 1 state to transition to the n = 3 or n = 5 or n = 7 states (*forbidden transitions*), whereas the n = 1 state *can* transition to the n = 2 or n = 4 or n = 6 states (*allowed transitions*).

A glow-in-the-dark watch provides an example of forbidden transitions. The paint on the hands absorbs light, but it contains "phosphorescent" material that cannot easily emit the light back out. It therefore stores the light. Ghostly green light leaks out slowly via a forbidden transition. The key is that the transition used to release the light is *almost* forbidden: it does happen, but slowly over a course of hours: it doesn't glow bright, but it glows long. Why does the forbidden transition occur at all? Perturbation theory is an approximation. The approximation is not perfect, hence the paint glows.

FIGURE 12.2 Phosphorescence. A "glow-in-the-dark" watch traps light energy. When the light is off, the energy slowly leaks away through "forbidden" transitions. [Photo by © VSchagow, https://commons.wikimedia.org/wiki/File:Regent-F997_frontal_Luminiszenz_(night-shot)_800_ISO_2022.jpg, licensed under CC BY-SA 4.0.]

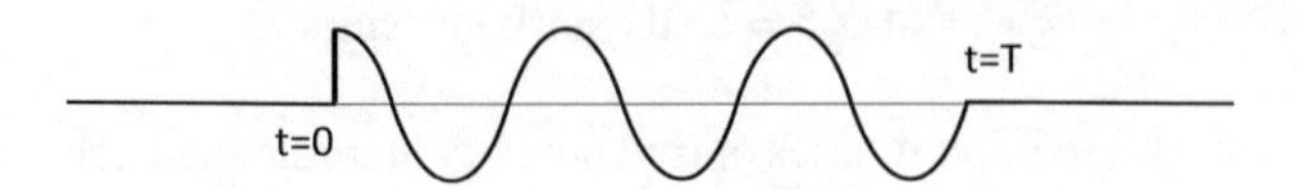

FIGURE 12.3 A time-dependent harmonic perturbation.

12.3 TDPT EXAMPLE 2: HARMONIC PERTURBATION (PERTURBATION OSCILLATES WITH TIME)

PROBLEM

Suppose that a particle is subject to an oscillating perturbation. For instance, light could be directed at the particle. Light waves, as you know, are an oscillating electromagnetic field. We need not be concerned with the nature of the perturbation here, only that it is turned on at t = 0, off at t = T, and it oscillates with frequency ω during that interval[1]:

$$\mathcal{H}_1(x,t) = 2\mathcal{H}_1(x)\cos(\omega t), \text{ active during } t = 0 \text{ to } T \qquad (12.7)$$

SOLUTION

Without any need to know the function $\mathcal{H}_1(x)$, the spatial variation of the perturbation, we can determine what sort of transitions are caused by the perturbation above (e.g., light of frequency ω). The transition amplitude is calculated from equation 12.2:

$$a_{ij} = \delta_{ij} - i/\hbar \int_0^T dt\, \mathcal{H}_{1ji}\, e^{i(\omega ji)t} = \delta_{ij} - i/\hbar \int_0^T dt\, 2\, \mathcal{H}_1(x)_{ji}\, e^{i(\omega ji)t} \cos(\omega t) \qquad (12.8)$$

where the $\mathcal{H}_1(x)$ and *cos(ωt)* parts of $\mathcal{H}_1$ have been separated[2]. Then $\mathcal{H}_1(x)_{ji}$ can be removed from the time integral. This leaves the following time integral:

$$\int_0^T dt\, e^{i(\omega ji)t}\, 2\cos(\omega t) \qquad (12.9)$$

This can be simplified by noting that $2\cos(\omega t) = e^{i\omega t} + e^{-i\omega t}$:

$$\int_0^T dt\, e^{i(\omega ji)t}[2\cos(\omega t)] = \int_0^T dt\, e^{i(\omega ji)t}[e^{i\omega t} + e^{-i\omega t}]$$
$$= \int_0^T dt\, [e^{i(\omega ji + \omega)t} + e^{i(\omega ji - \omega)t}]$$
$$= [e^{i(\omega ji + \omega)t} / i(\omega_{ji} + \omega) + e^{i(\omega ji - \omega)t} / i(\omega_{ji} - \omega)]_0^T$$
$$= \{e^{i(\omega ji + \omega)T} - 1\} / i(\omega_{ji} + \omega) + \{e^{i(\omega ji - \omega)T} - 1\} / i(\omega_{ji} - \omega) \qquad (12.10)$$

The key point is this: the 2nd term in equation 12.10 becomes very large when $\omega \sim \omega_{ji}$. The first term has no such point at which it gets as large, so it can be ignored.

The net result is:

When the frequency of the perturbation (ω) matches the frequency of transition ($\omega_{ji} = \Delta E/\hbar$), then the transition is highly probable.

If we sweep the frequency of light shining on the sample, various transitions will be caused, whenever the matching condition $\omega \sim \omega_{ji}$ occurs . This is the central principle of spectroscopy: the energy levels (more precisely, differences in energy levels) of a sample can be detected by noting which frequencies of light are strongly absorbed by the sample. Spectroscopy is explained via TDPT where the perturbation is light. The condition $\omega = \omega_{ji}$ is called the *resonance* condition, and thus absorption peaks in spectra are called resonances. When the light is off-resonance, absorption is relatively weak and can mostly be ignored.

SUMMARY

A PIB or square well is a good model system for particles (protons/neutrons) trapped in the nucleus by the strong force. So, for a nuclear-trapped proton, we've shown that the following perturbations cause the following effects:

constant ε field – nuclear E levels all change by the same amount while the field is on. Once the field is turned off, all energy levels return to their original values, but the distribution of particles amongst the levels has been altered. Only transitions by an odd number of levels are allowed: $\Delta n = +/- 1, 3, 5 \ldots$

oscillating perturbation – causes strong transitions if $\omega_{osc} \approx \Delta\omega_{states}$ *and* if $\mathcal{H}_1$ connects the states.

Note that we never used the eigenfunctions in discussing the oscillating perturbation. Therefore, the resonance result is general. Not just for nuclei (PIB), but for any sort of system, such as electrons in atoms or in molecules, or vibrational or rotational states of molecules, transitions can happen when the perturbation oscillation frequency matches the transition frequency. This is why the concept of resonance can be applied to all sorts of spectroscopy.

PROBLEMS

12.1 **(a)** Briefly describe TDPT example 1 from this chapter. Be sure to define the problem clearly. Do not show any math or equations, but describe the resulting conclusions. **(b)** Do the same for TDPT example 2 from this chapter.

12.2 A one-dimensional particle in a box is subject to a constant perturbation of $V = V_0$ within the box ($x = 0$ to $x = L$). There is no change in the perturbation with time. Calculate the change in energies due to this perturbation (see Chapter 11).

12.3 A one-dimensional particle in a box is subject to a constant perturbation of $V = V_0$ within the box ($x = 0$ to $x = L$), but only for a brief period of time T. At the end of time T, calculate the probability that a particle initially in the ground state would transition to the second excited state.

NOTES

1 Here ω is the frequency of the perturbing light, not $E_i/\hbar$ of the states, which we will call ω_i and ω_j. Note also that "2" is introduced into equation 12.7 merely to simplify the math that follows.

2 For brevity, $(\omega_{oj} - \omega_{oi})$ is abbreviated as ω_{ji}.

13 Variational Method

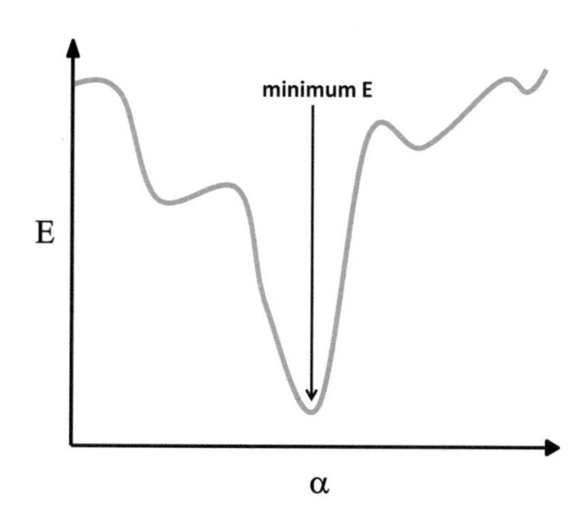

If the Hamiltonian does not closely resemble a simple Hamiltonian that we know how to solve, perturbation methods cannot be used. However, the Variational Method can be used to approximate the ground state eigenfunction and eigenenergy. No derivation will be given, only a recipe to follow to find the (approximate) solution, followed by an example.

13.1 VARIATIONAL METHOD RECIPE

1. $\psi_{ggs}(\alpha)$	guess at the form of the g.s. e.f., but include a built-in variable parameter α
2. $<E_{ggs}>$	calculate the expectation value of E using $\psi_{ggs}(\alpha)$
3. $\partial/\partial\alpha < E_{ggs}> = 0$	vary the parameter α so as to minimize the expectation value
4. $<E_{ggs}> \sim E_{gs}$	the resulting $<E_{ggs}>$ is an upper bound on the true g.s. E
5. $\psi_{ggs}(\alpha) \sim \psi_{gs}$	plug the minimizing value of α into $\psi_{ggs}(\alpha)$ to approximate the g.s. e.f.

13.2 EXAMPLE: SHO GROUND STATE

(For this section, assume we don't already know the exact answer.) There are few simple examples of use of the Variational Method. In studying atoms and molecules, Variational Method calculations are needed. However, they are too complicated to perform here. It is, though, important to know that this approximation can and has

DOI: 10.1201/9781003356172-16

been used by quantum chemists to find, for instance, a good approximation for the energy of each of the three electrons in a lithium atom.

We will use, as an example of the Variational Method, a particle in an SHO potential. Of course, we already know that for an SHO, $E_{gs} = ½ \hbar\omega_0$, and $\psi_{gs} \sim \exp(-\zeta^2/2)$, where ζ is related to x via equation 10.4. We will use these known results to check our variational results.

1. $\psi_{ggs}(\alpha)$	guess $\psi_{ggs} = \exp(-\alpha x^2)$ A good guess for ψ is an exponential function with *unknown* decay constant α
2. $< E_{ggs} >$	$< \exp(-\alpha x^2) \mid (-\hbar^2/2m)(\partial^2/\partial x^2) + ½ kx^2 \mid \exp(-\alpha x^2) >$ $= (\hbar^2/2m)\alpha + (k/8)(1/\alpha)$ The process of evaluating this integral is not shown. Requires normalization.
3. $\partial/\partial\alpha < E_{ggs} > = 0$	$\partial/\partial\alpha[(\hbar^2/2m)\alpha + (k/8)(1/\alpha)] = (\hbar^2/2m) + (-k/8)(1/\alpha^2) = 0$ $\rightarrow \alpha^2 = (k/8)(2m/\hbar^2)$, $\alpha = (1/2\hbar)\sqrt{(km)} = m\omega_o/2\hbar$ {since $\omega_o = \sqrt{(k/m)}$ for SHO}
4. $<E_{ggs}> \sim E_{gs}$	$(\hbar^2/2m)\alpha + (k/8)(1/\alpha)$ must be evaluated for $\alpha = m\omega_o/2\hbar$ multiplication and simplification yields $E_{gs} \sim (\hbar/2)\,\omega_o$
5. $\psi_{ggs}(\alpha) \sim \psi_{gs}$	plug the minimizing value of α into $\psi_{ggs}(\alpha)$ to approximate the true g.s. e.f.

Step 5 is left as an exercise for the reader (problem 13.1).

The Variational Method in this case finds the exact g.s. energy and ef, since our initial guessed wave function was a very good guess. In general, the results are only an approximation, and the approximation improves as the guess improves.

The excited states can also be approximated by the Variational Method, by requiring them to be orthogonal to the ground state, and to each other.

PROBLEMS

13.1 In the SHO example from this chapter, plug the calculated value of α (see step 3) into the guessed ground state wave function (see step 1) to find a good approximation for the true ground state wave function. Compare your answer to the SHO ground state in table 10.2.

13.2 When/why would the Variational Method be used instead of the perturbation method?

13.3 The electronic wave functions in a helium atom cannot be found exactly: approximation is required. In a helium atom, the electrons are attracted to the nucleus, as is the one electron in a hydrogen atom. However, there is the additional complication that the two electrons repel each other. **(a)** Is this repulsion a major or a minor consideration? **(b)** Given your answer for part (a), which approximation method should be used to account for electron-electron repulsion: perturbation theory, or the Variational Method?

Three-Dimensional Space: Atoms and Molecules

14 Generalization to 3D

14.1 THREE-DIMENSIONAL DERIVATIVES

To this point, we've assumed a particle existing in one dimension (x). We now discuss particles in three dimensions. How does this change the TISE? Simple, just replace $\partial/\partial x$ with ∇:

$$[-\hbar^2/2m(\partial/\partial x)^2 + V(x)]\,\psi = E\,\psi \;\rightarrow\; [-\hbar^2/2m\nabla^2 + V(\boldsymbol{r})]\,\psi = E\,\psi \quad (14.1)$$

where $V(\boldsymbol{r})$ represents the dependence of the potential on three dimensional coordinates. We do not wish to get bogged down in the formality of three-dimensional derivatives. We only need to know that ∇^2 represents a sum of double derivatives with respect to all three dimensions in space. These three dimensions could by x, y, and z. But since we are interested in electrons attracted to a nucleus at the center of an atom, we will use spherical coordinates r, θ and ϕ (see figure 14.1) which more naturally fit the situation. In spherical coordinates, ∇^2 takes the form:

$$\begin{aligned}\nabla^2 = {} & (1/r^2)\{(\partial/\partial r)\,[r^2\,(\partial/\partial r)]\} \\ & + (1/r^2)\{(1/\sin\theta)(\partial/\partial\theta)\,[\sin\theta\,(\partial/\partial\theta)]\} \\ & + (1/r^2)\{(1/\sin^2\theta)(\partial^2/\partial\phi^2)\} \\ = {} & (1/r^2)\{\underline{\boldsymbol{r}}\} + (1/r^2)\{\underline{\boldsymbol{\theta}}\} + (1/r^2)\{\underline{\boldsymbol{\phi}}\} \quad (14.2)\end{aligned}$$

where the three symbols $\underline{\boldsymbol{r}}$, $\underline{\boldsymbol{\theta}}$ and $\underline{\boldsymbol{\phi}}$ have been introduced to represent the three lengthy terms inside curly brackets above equation 14.2. These symbols help to simplify the discussion below.

One might have expected ∇^2 in spherical coordinates to be simply $\nabla^2 = (\partial/\partial r)^2 + (\partial/\partial\theta)^2 + (\partial/\partial\phi)^2$. These derivatives *are* present in equation 14.2, but additional factors of r and $\sin\theta$ also appear. This is reminiscent of the volume element dV required for three-dimensional integration using spherical coordinates. The spherical volume element is not simply $dr\,d\theta\,d\phi$, but is instead:

$$dV = r^2 \sin\theta\, dr\, d\theta\, d\phi \quad (14.3)$$

DOI: 10.1201/9781003356172-18

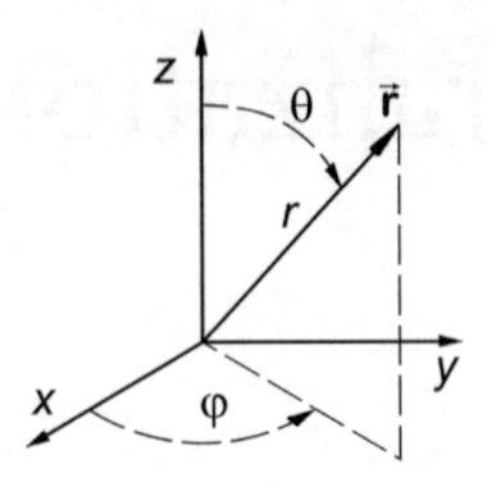

FIGURE 14.1 Spherical coordinates.

[Image by © P.wormer, https://commons.wikimedia.org/wiki/File:Spherical_polar_coordinates.png, licensed under CC BY-SA 3.0.]

The unique characteristics of spherical coordinates that cause factors of r and $sin\theta$ to appear in the volume element also cause them to appear in ∇^2.

Substituting the spherical version of ∇^2 from equation 14.2 into the TISE (equation 14.1) leads to:

TISE (spherical coordinates):

$$(-\hbar^2/2m)\ (1/r^2)\ \{\underline{\boldsymbol{r}} + \underline{\boldsymbol{\theta}} + \underline{\boldsymbol{\phi}}\}\ \psi + V(\boldsymbol{r})\psi = E\psi \qquad (14.4)$$

Recall that the TISE is just $KE + V = E$. In equation 14.4, the three terms on the left each contain two derivatives with respect to r, θ or ϕ, respectively. These terms must therefore represent KE in the r, θ and ϕ directions, respectively.

The solutions (ψ) to the TISE (equation 14.1) are thus functions of r, θ, and ϕ. Clearly, this is a complicated differential equation. To solve it, first isolate the derivative terms:

$$\underline{\boldsymbol{r}}\psi + \underline{\boldsymbol{\theta}}\psi + \underline{\boldsymbol{\phi}}\psi = (2mr^2/\ \hbar^2)[V(\boldsymbol{r}) - E]\ \psi \qquad (14.5)$$

Equation 14.5 is still complicated: the left-hand side (LHS) contains derivatives with respect to three variables (r, θ, ϕ), and the right-hand side (RHS) has a potential $V(\boldsymbol{r})$ that could depend on the same three variables. To simplify, we assume a potential depending only on r, which is appropriate for an electron orbiting a nucleus: its PE depends only on the distance from the nucleus, not on whether it is above/below, or to the right/left of the nucleus; in symbols, $V = V(r)$. In this case, it becomes possible to separate the r variable from the two angular variables, similar to how x and t were separated in Chapter 5, when V depended only upon x. Separation will simplify analysis.

To separate r from θ and ϕ, we follow a procedure analogous to section 5.3. First, assume that a separable solution can be found such that $\psi(r, \theta, \phi)$ can be written as a radial function $R(r)$ times an angular function $Y(\theta, \phi)$:

$$\psi(r, \theta, \phi) = R(r)\ Y(\theta, \phi) \qquad (14.6)$$

Inserting this form of $\psi(r, \theta, \phi)$ into equation 14.5 yields:

$$\underline{\boldsymbol{r}}RY + \underline{\boldsymbol{\theta}}RY + \underline{\boldsymbol{\phi}}RY = (2mr^2/\hbar^2)[V(r)-E]\,RY \tag{14.7}$$

Next, divide both sides by $\psi = RY$ and cancel terms not acted on by operators:

$$(\underline{\boldsymbol{r}}R)/R + (\underline{\boldsymbol{\theta}}Y)/Y + (\underline{\boldsymbol{\phi}}Y)/Y = (2mr^2/\hbar^2)[V(r)-E] \tag{14.8}$$

Isolate the *r* terms to the LHS:

$$(\underline{\boldsymbol{r}}R)/R + (2mr^2/\hbar^2)[E - V(r)] = -(\underline{\boldsymbol{\theta}}Y)/Y - (\underline{\boldsymbol{\phi}}Y)/Y \tag{14.9}$$

The LHS of equation 14.9 is a function only of *r*, while the RHS is a function only of θ and ϕ. Following the logic of section 5.4, we conclude that both sides equal a constant. We will call this separation constant $\boldsymbol{\ell(\ell+1)}$, because it will be handy later.

Separating the two sides of the equation, setting each equal to $\boldsymbol{\ell(\ell+1)}$, and multiplying the *r* equation by *R* and the angular equation by *Y*, produces the radial TISE and the angular TISE:

$$(\underline{\boldsymbol{r}}R) + (2mr^2/\hbar^2)[E - V(r)]\,R = \boldsymbol{\ell(\ell+1)}\,R \quad \text{radial TISE} \tag{14.10}$$

$$(\underline{\boldsymbol{\theta}}Y) + (\underline{\boldsymbol{\phi}}Y) = -\boldsymbol{\ell(\ell+1)}\,Y \quad \text{angular TISE} \tag{14.11}$$

Apparently, the solutions *R(r)* to the radial TISE depend on the exact form of *V(r)*. It cannot be solved until *V(r)* is defined. That is, the solutions *R(r)* differ for each different form of *V(r)*.

However, *V(r)* does *not* appear in the angular TISE equation, so its solutions $Y(\theta, \phi)$ do *not* depend on the form of *V(r)*: they are general solutions for *any* radial potential. They are applicable for an electron bound to a hydrogen nucleus or, for instance, for the earth orbiting the sun. We need only solve the angular TISE once (just as we solved the SISE only once). The solutions will fit *all* radially symmetric potentials that we encounter. We now proceed to solve the angular TISE.

14.2 ANGULAR TISE

$$\underline{\boldsymbol{\theta}}Y + \underline{\boldsymbol{\phi}}Y = -\,\boldsymbol{\ell(\ell+1)}\,Y \tag{14.12}$$

The angular TISE is still not simple, since two variables (θ, ϕ) are involved. Can these two variables be separated? Try by assuming the separable solution $Y(\theta, \phi) = P(\theta)F(\phi)$, and plugging it into equation 14.12, while restoring the long version of the derivative operators:

$$\{(1/\sin\theta)(\partial/\partial\theta)\,[\sin\theta\,(\partial/\partial\theta)]\}\,PF + \{(1/\sin^2\theta)(\partial^2/\partial\phi^2)\}\,PF = -\,\boldsymbol{\ell(\ell+1)}\,PF \tag{14.13}$$

Next, divide both sides by $Y = PF$, and cancel terms not acted on by operators:

$$(1/P)\{(1/\sin\theta)(\partial/\partial\theta)\,[\sin\theta\,(\partial/\partial\theta)]\}P + (1/F)\{(1/\sin^2\theta)(\partial^2/\partial\phi^2)\}F = -\boldsymbol{\ell}(\boldsymbol{\ell}+\boldsymbol{1}) \quad (14.14)$$

The two terms on the LHS are *almost* separated into a θ and a ϕ term, but the 2nd term has a factor of $1/\sin^2\theta$. Remove it by multiplying both sides by $\sin^2\theta$:

$$(1/P)\{(\sin\theta)(\partial/\partial\theta)\,[\sin\theta\,(\partial/\partial\theta)]\}P + (1/F)\{(\partial^2/\partial\phi^2)\}F = -\boldsymbol{\ell}(\boldsymbol{\ell}+\boldsymbol{1})\sin^2\theta \quad (14.15)$$

Now isolate the ϕ term to the RHS:

$$(1/P)\{(\sin\theta)(\partial/\partial\theta)\,[\sin\theta\,(\partial/\partial\theta)]\}P + \boldsymbol{\ell}(\boldsymbol{\ell}+\boldsymbol{1})\sin^2\theta = -(1/F)\{(\partial^2/\partial\phi^2)\}F \quad (14.16)$$

The LHS is a function of only θ, the RHS is a function only of ϕ. This requires that each side equals a constant. We will call this separation constant $\boldsymbol{m}^2$, because it will be handy later. We now have two equations:

$$(1/P)\{(\sin\theta)(\partial/\partial\theta)\,[\sin\theta\,(\partial/\partial\theta)]\}P + \boldsymbol{\ell}(\boldsymbol{\ell}+\boldsymbol{1})\sin^2\theta = \boldsymbol{m}^2 \quad (14.17)$$

$$-(1/F)\{(\partial^2/\partial\phi^2)\}F = \boldsymbol{m}^2 \quad (14.18)$$

These can be rearranged to resemble eigenfunction equations, by multiplying the θ equation by P and the ϕ equation by F, to form the θ TISE and the ϕ TISE:

$$\{(\sin\theta)(\partial/\partial\theta)\,[\sin\theta\,(\partial/\partial\theta)]\}P = [\boldsymbol{m}^2 - \boldsymbol{\ell}(\boldsymbol{\ell}+\boldsymbol{1})\sin^2\theta]\,P \qquad \theta \text{ TISE} \quad (14.19)$$

$$(\partial^2/\partial\phi^2)F = -\boldsymbol{m}^2\,F \qquad \phi \text{ TISE} \quad (14.20)$$

We now seek solutions $P(\theta)$ and $F(\phi)$ that solve these two equations. The ϕ TISE is trivial to solve. We at this point choose only one of the possible solutions:

$$F(\phi) = e^{+im\phi} \quad (14.21)$$

Next, we apply general conditions (see section 6.1) to $F(\phi)$. Specifically, $F(\phi)$ must be single-valued. That is, F cannot have two different values for one value of ϕ. Since $\phi = 0$ and $\phi = 2\pi$ are the same angle, $F(\phi)$ must equal $F(\phi + 2\pi)$. Problem 14.1 shows that this condition requires $\boldsymbol{m}$ to be an integer:

$$\boldsymbol{m} = 0, \pm 1, \pm 2, \pm 3 \ldots . \quad (14.22)$$

The θ TISE (equation 14.15) is more complicated. Its derivatives are separated by a factor of $\sin\theta$, and it contains both $\boldsymbol{m}$ and $\boldsymbol{\ell}$. In fact, the θ TISE is actually a *series* of differential equations, one for each combination of $\boldsymbol{m}$ and $\boldsymbol{\ell}$. Each equation must be solved separately. Furthermore, we already know that each $\boldsymbol{m}$ will be an integer. As a result, it turns out to be impossible to find solutions P(θ) that are single-valued and finite, unless ℓ is a whole number and the magnitude of ℓ is at least equal to the magnitude of m . In other words, solutions P(θ) exist only if:

$$\ell \text{ is a whole number such that } |m| <= \ell \quad (14.23)$$

This restriction is more typically stated as follows:

$$\ell = 0,1,2 \ldots \quad (14.24)$$

$$m = -\ell \text{ to } +\ell \text{ (every integer from } -\ell \text{ to } +\ell\text{)} \quad (14.25)$$

The solutions are called the associated Legendre polynomials $P_{\ell m}(\theta)$ (table 14.1, first column). Multiplication of $P_{\ell m}(\theta)$ by $F(\phi)$ yields the angular eigenfunctions $Y_{\ell m}(\theta, \phi)$, also called the spherical harmonics in complex form (table 14.1, second column).

$$Y_{\ell m}(\theta, \phi) = P_{\ell m}(\theta)F(\phi) \quad (14.26)$$

{spherical harmonic = associated Legendre polynomial multiplied by $e^{im\phi}$}

TABLE 14.1
Spherical harmonics

Associated Legendre Polynomials P_ℓ^m	Complex Spherical Harmonics Y_ℓ^m	Real Spherical Harmonics
$P_0^0 = 1$	$Y_0^0 = 1$	$Y_s = 1$
$P_1^0 = \cos\theta$	$Y_1^0 = \cos\theta$	$Y_{pz} = \cos\theta$
$P_1^{\pm 1} = \sin\theta$	$Y_1^{\pm 1} = \sin\theta\ e^{\pm i\phi}$ →	$Y_{px} = \sin\theta\ \cos(\phi)$
		$Y_{py} = \sin\theta\ \sin(\phi)$
$P_2^0 = ½(3\cos^2\theta - 1)$	$Y_2^0 = ½(3\cos^2\theta - 1)$	$Y_{dz2} = ½(3\cos^2\theta - 1)$
$P_2^{\pm 1} = 3\cos\theta\ \sin\theta$	$Y_2^{\pm 1} = 3\cos\theta\ \sin\theta\ e^{\pm i\phi}$ →	$Y_{dxz} = 3\cos\theta\ \sin\theta\ \cos(\phi)$
$P_2^{\pm 2} = 3\sin^2\theta$	$Y_2^{\pm 2} = 3\sin^2\theta\ e^{\pm 2i\phi}$ →	$Y_{dyz} = 3\cos\theta\ \sin\theta\ \sin(\phi)$
		$Y_{dx2-y2} = 3\sin^2\theta\ \cos(2\phi)$
		$Y_{dxy} = 3\sin^2\theta\ \sin(2\phi)$

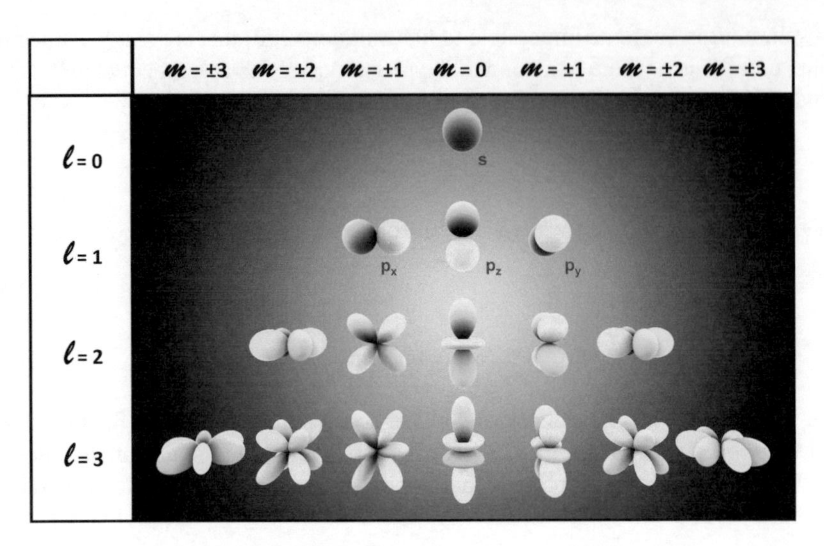

FIGURE 14.2 The *real* spherical harmonics. Blue and yellow lobes represent positive and negative density, respectively. [Image by © Inigo.quilez, https://commons.wikimedia.org/wiki/File:Spherical_Harmonics.png, licensed under CC BY-SA 3.0.]

These complex functions can be rearranged into real functions, as discussed in Chapter 15. The results are presented in the last column of table 14.1.

Figure 14.2 presents polar plots for the real spherical harmonics. You will recognize that these resemble electron orbitals. These plots indicate the direction in space that an electron is likely to be found, but *not* how far away from the nucleus the electron will be. For distance information, the functions $R(r)$ that solve the radial TISE (equation 14.10) are needed. That is the subject of Chapter 16. But first, in Chapter 15, the angular TISE solutions (spherical harmonics) are discussed in terms of their angular momentum. It should not surprise you to discover that a larger angular momentum is associated with a larger number of nodes (nodal planes).

PROBLEMS

14.1 Use the Euler identity to show that for the function $F(\phi) = e^{+im\phi}$, integer values of $\boldsymbol{m}$ guarantee that $F(\phi) = F(\phi + 2\pi)$.

14.2 **(a)** Prove that the associated Legendre polynomial $P_2^{\,1}$ satisfies the θ TISE (equation 14.15A).
(b) Prove that the function $e^{+i\phi}$ satisfies the ϕ TISE (equation 14.16).
(c) Prove that the spherical harmonic $Y_2^{\,1}$ satisfies the angular TISE (eq. 14.11).

14.3 Name a type of electron orbital related to each of the images in figure 14.2. Specify the values of $\boldsymbol{\ell}$ and $\boldsymbol{m}$ in each case. (Hint: the quantum number $\boldsymbol{n}$ has not yet been specified.)

14.4 Using figure 14.2, count the number of nodal planes in the following spherical harmonic functions: **(a)** the $\ell = 0$; **(b)** each $\ell = 1$; **(c)** each $\ell = 2$; **(d)** each $\ell = 3$. Use these results to propose a simple formula for the number of angular nodes in a spherical harmonic.

14.5 Using table 14.1, derive a relationship between the value of ℓ and the power of the trigonometric functions in the associated Legendre polynomials.

14.6 In table 14.1, why are P_2^0 and Y_2^0 identical, whereas P_2^1 and Y_2^1 differ?

15 Angular Momentum

15.1 QUANTIZATION OF ORBITAL ANGULAR MOMENTUM (L)

The angular TISE (equation 14.12) is:

$$\underline{\theta}Y + \underline{\phi}Y = -\ell(\ell + 1)\,Y \tag{15.1}$$

Multiplication of equation 15.1 by $(-\hbar^2/2m)\,(1/r^2)$ returns it to a form of the angular TISE that more closely resembles the original 3D TISE (equation 14.1 or equation 14.4):

$$(-\hbar^2/2mr^2)\{\underline{\theta}Y + \underline{\phi}Y\} = (\hbar^2\,\ell(\ell + 1)\,/2mr^2)\,Y \tag{15.2}$$

Recall (see discussion below equation 14.4) that the two terms on the LHS are the KE in the θ and ϕ directions, respectively. Thus, together, they are the total angular KE. In classical mechanics, just as linear KE = $p^2/2m$, angular KE = $L^2/2I$, where L is angular momentum and I is the moment of inertia, which for a point mass simply equals mr^2. Note the factors of $I = 2mr^2$ on both sides of equation 15.2. Apparently equation 15.2 simply states: angular KE = $L^2/2I$.

So, equation 15.2 is actually an eigenvalue equation for angular KE. It can be converted into an eigenvalue equation for L^2 (= KE*2I) by multiplying both sides by $2I = 2mr^2$:

$$-\hbar^2\{\underline{\theta}Y + \underline{\phi}Y\} = \hbar^2\,\ell(\ell + 1)\,Y \tag{15.3}$$

Thus, equation 15.3 can more succinctly be written in the form:

$$\hat{L}^2 Y = L^2 Y \tag{15.4}$$

where the LHS contains the $\hat{L}^2$ operator:

$$\hat{L}^2 = -\hbar^2\left\{\underline{\theta} + \underline{\phi}\right\} \tag{15.5}$$

DOI: 10.1201/9781003356172-19

which, when applied to the eigenfunction Y, gives the same function back times the L^2 eigenvalue:

$$L^2 = \hbar^2 \ell(\ell + 1) \tag{15.6}$$

or

$$L = \hbar\sqrt{\ell(\ell + 1)} \tag{15.7}$$

FIGURE 15.1 Angular momentum. An ice skater spinning counterclockwise about his axis has angular momentum in the z-direction (up). [Photo by © Carolin Paré, https://commons.wikimedia.org/wiki/File:Karel_Zelenka_Spin_-_2007_Europeans.jpg, licensed under CC BY-SA 3.0. Arrows were added]

What we have shown is that for *any* quantum particle subject to *any* central potential, the angular momentum L is quantized to the values given by equations 15.6 and 15.7, where ℓ can take on only whole number values (ℓ = 0, 1, 2, 3). Thus, so long as the particle remains in the state represented by a particular spherical harmonic function $Y_{lm}(\theta,\phi)$, its total angular momentum is conserved at the value given by equation 15.7. This matches the classical principle that angular momentum is conserved, unless an angular force is applied. But it goes beyond that: it specifies not only *conservation*, but also *quantization* of the total angular momentum.

Furthermore, just as the linear momentum operator is $\hat{p} = \hbar/i(\partial/\partial x)$, it can be shown that the $\hat{L}_z$ operator, representing angular momentum around the z axis, is:

$$\hat{L}_z = \hbar/i\left(\partial/\partial\phi\right) \tag{15.8}$$

To see whether the spherical harmonics are eigenfunctions of $\hat{L}_z$, simply operate on them with the $\hat{L}_z$ operator, and test whether the result is the same function times a constant:

$$\hat{L}_z P(\theta)F(\phi) = P(\theta)\hbar/i\left(\partial/\partial\phi\right)e^{im\phi} = \boldsymbol{m}\hbar\, P(\theta)e^{im\phi} \tag{15.9}$$

where the form $Y = P(\theta)F(\phi)$ was used (see section 14.2) and $P(\theta)$was passed through the derivative since it does not contain ϕ. In fact, the entire equation 15.9 can be divided by $P(\theta)$ to yield:

$$\hat{L}_z F(\phi) = \hbar/i\left(\partial/\partial\phi\right)e^{im\phi} = \boldsymbol{m}\hbar\, e^{im\phi} \tag{15.10}$$

Apparently the spherical harmonics $Y_{lm}(\theta,\phi) = P_{lm}(\theta)F(\phi)$ are eigenfunctions of $\hat{L}_z$, but only because $F(\phi) = e^{im\phi}$ are eigenfunctions of $\hat{L}_z$. $P(\theta)$ plays no role. The L_z eigenvalues are $\boldsymbol{m}\hbar$. The angular momentum about the z axis is therefore also quantized in integral units of ħ, and L_z is also conserved[1].

Any particle subject to a central potential V(r) has:

- conserved, quantized angular momentum of $L = \hbar\sqrt{\ell(\ell+1)}$
- conserved, quantized angular momentum around the z-axis of $L_z = \boldsymbol{m}\hbar$
- where $\boldsymbol{m} = -\ell$ to $+\ell$

15.2 ALTERNATE FORMS OF F(ϕ): EXPONENTIAL AND TRIGONOMETRIC

You may recognize that ℓ is the orbital quantum number, while $\boldsymbol{m}$ is the magnetic quantum number. Note that both quantum numbers arise as separation variables: $\ell(\ell+1)$ was used to separate the radial and angular equation, and $\boldsymbol{m}^2$ was used to separate the θ and ϕ variables. $\boldsymbol{E}$ was a separation constant used to separate x and t (see Chapter 5).

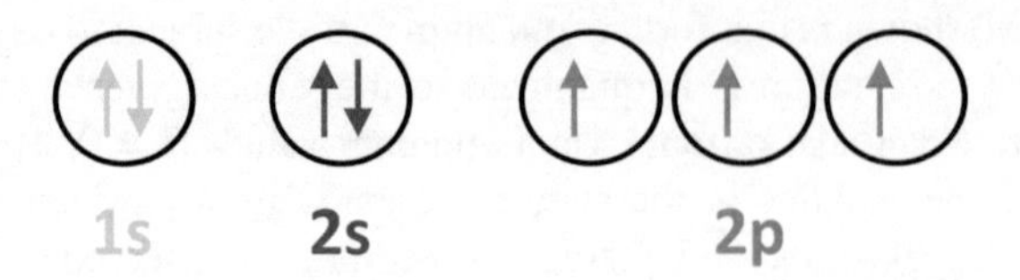

FIGURE 15.2 Nitrogen orbitals. The ground electronic state of nitrogen has one electron in each of the 2p orbitals p_x, p_y and p_z. Each p-orbital corresponds to $\ell = 1$. What is the value of $\boldsymbol{m}$ for each of these three orbitals?

The spherical harmonics Y_{lm} (θ, ϕ) are the angular part of the wave function for any particle subject to a central potential, including electrons attracted to a nucleus in an atom. That is why we are interested in the spherical harmonics. First year chemistry students are taught to fill electrons into orbitals with diagrams such as that shown in figure 15.2. They are also told that the three *2p* boxes represent p_x, p_y, and p_z orbitals. Now we have derived the quantum numbers $\boldsymbol{\ell}$ and $\boldsymbol{m}$, and we recognize that *p* orbitals have $\boldsymbol{\ell} = 1$, and therefore $\boldsymbol{m} = -1, 0, +1$. Which *2p* orbital in figure 15.2 represents which value of $\boldsymbol{m}$?

This is a trick question. The bottom line (see problem 15.1) is that the middle circle can be labeled as $\boldsymbol{m} = 0$ and identified as the p_z orbital. However, the p_x and p_y orbitals are each a linear combination of $e^{+im\phi}$ with $\boldsymbol{m} = 1$ and $\boldsymbol{m} = -1$:

$$p_x = \cos(\phi) = (1/2)(e^{+i\phi} + e^{-i\phi}) \quad p_y = \sin(\phi) = (1/2i)(e^{+i\phi} - e^{-i\phi}) \qquad (15.11)$$

It is easily shown that the functions $\cos(\boldsymbol{m}\phi)$ and $\sin(\boldsymbol{m}\phi)$ are also solutions to the ϕ TISE (14.16). In fact, we are free to choose either $\cos(\boldsymbol{m}\phi)$ and $\sin(\boldsymbol{m}\phi)$, or $e^{\pm im\phi}$ as the TISE eigenfunctions. This situation is somewhat analogous to the fact that electrons can be thought of as particles or as waves. We are free to choose the representation that helps us answer a particular problem.

In this case, if we wish to specify the value of $\boldsymbol{m}$, then $e^{im\phi}$ is the better choice. However, if we desire *p* orbitals pointing in specific directions, the $\cos(\boldsymbol{m}\phi)$ and $\sin(\boldsymbol{m}\phi)$ solutions are preferred: they point along x or along y, respectively. For chemists interested in bonding, the trigonometric functions are thus preferred. But this is not sorcery: each form is merely a linear combination of the other form: there is no prohibition against using a linear combination of eigenfunctions to create another solution.

The main point is that we will use the trigonometric forms of $F(\phi)$ from this point forward. We will bear in mind that they represent linear combinations of the exponential forms, so that the p_x and p_y orbitals actually have an undetermined $\boldsymbol{m}$ value: each has a mixture of $\boldsymbol{m} = \pm 1$ (see table 14.1 and figure 14.2).

15.3 SPIN ANGULAR MOMENTUM (S)

It was discussed in Chapter 14 that, since the earth is subject to the central potential of the sun, the earth's orbital angular momentum about the sun is conserved

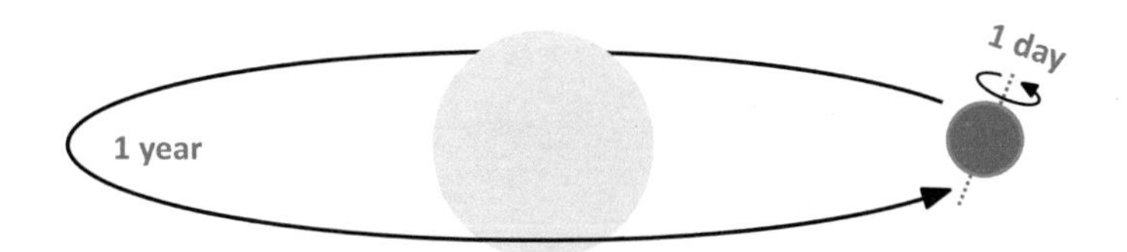

FIGURE 15.3 Orbital and spin angular momentum. Just as the earth orbits around the sun *and* spins about its own axis, an electron also behaves as though it orbits around the nucleus and spins about its own axis.

and quantized. But the earth also spins on its axis. That is also a form of angular momentum, and it is also conserved and quantized.

The same is true for an electron "orbiting" a nucleus. While it orbits the nucleus with a conserved quantized angular momentum L, it also spins on its axis with a separate conserved quantized angular momentum we call S, which stands for spin angular momentum. Though there are similarities between L and S, there are also differences. First, conceptually, L denotes translational motion of the electron around the nucleus, while S describes a rotation of the electron about its own center. Therefore, we need not describe "orbitals" for the spin in order to visualize the associated motion. We only need the quantized values of S, which basically means how fast it is spinning. Apparently, quantum particles can only spin at certain speeds. We won't calculate the speed, just the angular momentum associated with spinning: that is more directly related to energy.

Note that no-one has ever "seen" an electron spinning. In fact, it may not spin, but it behaves as though it does. That is, measurements made on electrons (and other particles) can be understood if it is assumed that electrons are spinning with the angular momenta that we are about to calculate. Calling this "spin" is just a convenient visual concept.

The quantized values of S (and S_z) are much like their orbital counterparts L and L_z, except that half-integer values can be used. That is, while the quantum numbers associated with L are restricted to whole numbers ($\boldsymbol{\ell}$) or integers ($\boldsymbol{m_\ell}$), the analogous quantum numbers associated with S can, in general, be positive half-integers (**s**) and positive or negative half-integers ($\boldsymbol{m_s}$) in addition to the whole number and integer values. How can this be possible?

Recall we determined that $\boldsymbol{m_l}$ must be an integer by using the boundary condition that $F(\phi) = F(\phi + 2\pi)$. There should be a similar function associated with S, but if half-integers are allowed for $\boldsymbol{m_s}$, then apparently the boundary condition is $F(\phi) = F(\phi + 4\pi)$. There is no way to justify this change in boundary condition at this time. The half-integer values of these quantum numbers were discovered experimentally before they were understood theoretically.

To derive this fact requires the use of relativistic quantum mechanics, which is well beyond our scope. We will just use the result: integral and half-integral values of quantum numbers are allowed for S. These values can then be substituted into formulae analogous to equations 15.6 and 15.7 to calculate the angular momentum associated with spin (we call this S rather than L).

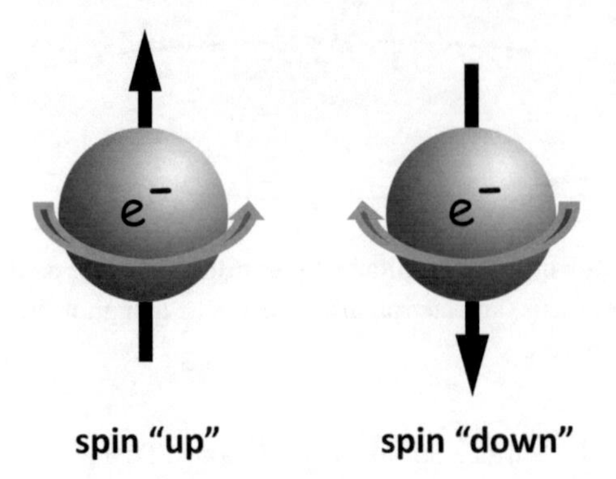

FIGURE 15.4 Electron spin states. An electron is a spin-½ particle, meaning the quantum number $\boldsymbol{s} = ½$. The quantum number $\boldsymbol{m_s}$ therefore can be +½ (spin up) or –½ (spin down). The spin rate is equal for both states, but the direction of spinning is opposite.

Briefly, the justification for the existence of half-integral spin is that the TISE is not relativistic: it treats time and space differently (two space derivatives, one time derivative). In order to make the equation relativistic, either both derivatives should be double (Klein-Gordon equation) or both derivatives should be single (Dirac equation). The Dirac equation turns out to be the correct form, and its treatment leads to the derivation of spin and its allowed half-integral quantum numbers.

It turns out that electrons are *fermions*, which is simply a name given to particles with *half-integral* spin. Protons and neutrons are also fermions. Photons (light particles) are an example of *bosons*, which is the name given to particles with *integral* spin. We are interested in electrons, so focusing on their case, the electron spin quantum number **s** always equals ½. That is, all electrons "spin" at the same speed. The only spin degree of freedom for the electron is in its 2nd quantum number $\boldsymbol{m_s}$, which ranges from -**s** to +**s** by integral steps (analogous to equation 14.21). Thus, $\boldsymbol{m_s}$ can only equal +½ (spin "up") or –½ (spin "down"). So while the electron spins with only one speed, observing the spinning about the z-axis would show that it spins clockwise ($\boldsymbol{m_s}$ = +½) or counterclockwise ($\boldsymbol{m_s}$ = –½). The "spin up" and "spin down" states could instead be called "spin clockwise" and "spin counterclockwise" (figure 15.4)

PROBLEMS

15.1 For the functions $F(\phi) = sin(\boldsymbol{m}\phi)$ and $F(\phi) = cos(\boldsymbol{m}\phi)$: **(a)** show that each solves the ϕ TISE; **(b)** show that neither is an ef of $\hat{L}_z$; **(c)** explain/discuss your answer to part (b); and **(d)** find $\langle L_z \rangle$ for each function.

15.2 List the quantum numbers associated with the following orbitals **(a)** $4p_x$ **(b)** $4p_y$ **(c)** $4p_z$.

15.3 Calculate the values of L for each of the orbitals in problem 15.2. Provide your answers in terms of multiples of ħ.

15.4 Calculate the values of L_z for each of the orbitals in problem 15.2. Leave your answers in terms of multiples of ħ. If the values of L_z are not unique, then state that and provide instead $\langle L_z \rangle$.

15.5 Calculate the values of S for an electron. Leave your answers in terms of multiples of ħ.

15.6 Calculate the values of S_z for an electron in **(a)** the spin up state; **(b)** the spin down state. Leave your answers in terms of multiples of ħ.

15.7 Deuterium (^2H) is a spin 1 nucleus. **(a)** Compare its spin rate to the spin rate of an electron. **(b)** List the possible values of m_s. **(c)** Discuss/interpret these values of m_s: what do they mean?

15.8 For the nucleus ^{23}Na, which has a spin quantum number s = 3/2: **(a)** calculate the spin angular momentum **S**; **(b)** list all possible m_s values; and **(c)** calculate all possible S_z values.

NOTE

1 It is quite astounding how often Planck's constant pops up in quantum mechanics.

16 H-Atom
Solving the Radial TISE

16.1 RADIAL EQUATION: H-ATOM ELECTRON

After separation of variables, the radial TISE, as given in equation 14.10, is

$$(\underline{\boldsymbol{r}}R) + (2mr^2/\hbar^2)[E + ke^2/r]\,R = \boldsymbol{\ell}(\boldsymbol{\ell} + \boldsymbol{1})\,R \quad (16.1)$$

where

$$\underline{\boldsymbol{r}} = \{(\partial/\partial r)\,[r^2\,(\partial/\partial r)]\} \quad (16.2)$$

and since our focus is the hydrogen atom, the potential V(r) has been explicitly written as

$$V(r) = -ke^2/r \quad (16.3)$$

We already know that the orbital quantum numbers $\boldsymbol{\ell}$ appearing in equation 16.1 must be whole numbers, and that they represent angular momentum. The $\boldsymbol{\ell}\,(\boldsymbol{\ell} + 1)$ term in equation 16.1 can be thought of as a "pseudopotential" in the radial dimension. The electron is attracted to the positively charged nucleus, but why does it not fall into the nucleus? Because it has angular momentum. The pseudopotential term accounts for the angular momentum that keeps the electron away from the nucleus.

Our task is to find solutions R(r) that satisfy equation 16.1. This is difficult. We first outline what the solutions should look like, and then list them in a table. First, note that solutions should decay to zero as $r \rightarrow \infty$: the electron should have vanishing probability to be found infinitely far from the nucleus. Therefore, an exponential factor is expected. This factor should decay rapidly for low energy electrons, and more slowly for high energies, to allow them to wander further from the nucleus. We might also expect a polynomial factor, as we found for the SHO, which would produce variations with r beyond a simple decay function, and introduce *maxima* and *nodes* at specific distances from the nucleus.[1]

DOI: 10.1201/9781003356172-20

16.2 FOUR RADIAL EQUATIONS (ℓ = 0,1,2,3)

Equation 16.1 is actually an infinite series of equations, one for each value of ℓ. As chemists, we need only be concerned with $\ell = 0, 1, 2, 3$ which correspond to s, p, d, and f orbitals, respectively.

Let's start with the $\ell = 0$ equation, which makes the RHS of equation 16.1 equal zero. A series of solutions can be found for the $\ell = 0$ case, the simplest three of which are shown in table 16.1. We label the simplest solution as $n = 1$, the next simplest as $n = 2$, the third as $n = 3$. Note that each solution contains an exponential factor, but this factor differs for the three solutions: n appears in the denominator of each exponent. Each $\ell = 0$ solution also contains a polynomial factor of order $n - 1$.

We look next at the case $\ell = 1$, for which two solutions are shown in table 16.1. Note that the simplest $\ell = 1$ solution looks a lot like the $n = 2, \ell = 0$ solution: it has a 2 in the exponent, and a first order polynomial. Therefore, although it is the "first" $\ell = 1$ solution, it makes sense to label it as $n = 2, \ell = 1$, and group it with the $n = 2, \ell = 0$ solution in table 16.1. Likewise the "second" $\ell = 1$ solution looks a lot like the $n = 3, \ell = 0$ solution, and so we call it the $n = 3, \ell = 1$ solution.

This process is repeated for the case $\ell = 2$. Here the first solution looks like both the $n = 3, \ell = 0$ and $n = 3, \ell = 1$ solutions, so we label it as the $n = 3, \ell = 2$ solution. Though not shown in the table, you should be able to predict what the first solution for the case $\ell = 4$ looks like. You may recognize that these results are the basis for why p, d, and f orbitals occur only for shells (n values) of at least 2, 3, and 4, respectively.

The number n, used to number the solutions R_{nl}, is called the *principal quantum number*. It can be shown that no solution to the radial TISE can be found unless $n > \ell$, so that, for each value of n, there are n different solutions corresponding to $\ell = n - 1, n - 2 \ldots 0$. For instance, for $n = 3$, there are three solutions R_{nl}: R_{32}, R_{31}, R_{30}. These are the radial parts of the 3d, 3p, and 3s orbitals, respectively.

16.3 RADIAL DISTANCE AND RADIAL NODES

Let's inspect the solutions R_{nl} in more detail (table 16.1). Each solution decays to zero as $r \rightarrow \infty$, but lower energy (lower n) solutions decay more rapidly. Thus, as expected, lower shell electrons are found closer to the nucleus.

Higher shell (higher n) solutions contain higher order polynomials. Therefore, they *can* have more nodes. The bottom line is that an s orbital, for instance, is not a simple sphere with the electron inside. Even a 1s orbital indicates some probability for the electron to be found at a large distance from the nucleus. What's more, a 2s orbital is even more complicated, and contains a radial node and two radial maxima. The 3s orbital contains two radial nodes and three radial maxima. The orbitals are far more complicated than assumed in a first-year chemistry class.

The radial functions for p, d, and f orbitals also *can* contain nodes. Of the p-orbitals, only the 2p orbital has no radial node, the 3p has one, and the 4p has two, etc. Likewise, the 3d orbital has no radial nodes, 4d has one, 5d has two, etc. From these examples, the pattern should be clear (see figure 16.1): the first shell in which an orbital type (s, p, d, f ...) appears produces no radial nodes for that orbital, and the number of nodes increases with each increase in the shell number (n).

TABLE 16.1
Radial wave functions R_{nl} and their related spherical harmonic functions. The indicated product (arrows) of radial functions times angular functions produces the atomic orbitals

n	ℓ	name	$R_{n\ell}$
1	0	R_{10}	e^{-r/a_o}
2	0	R_{20}	$[2 - r/a_o]\, e^{-r/2a_o}$
2	1	R_{21}	$[r/a_o]\, e^{-r/2a_o}$
3	0	R_{30}	$[27 - 18\,(r/a_o) + 2\,(r/a_o)^2]\, e^{-r/3a_o}$
3	1	R_{31}	$[6\,(r/a_o) - (r/a_o)^2]\, e^{-r/3a_o}$
3	2	R_{32}	$[(r/a_o)^2]\, e^{-r/3a_o}$

$Y_{pz} = \cos\theta$

$Y_{px} = \sin\theta\, \cos(\phi)$

$Y_{py} = \sin\theta\, \sin(\phi)$

$Y_{dz2} = ½(3\cos^2\theta - 1)$

$Y_{dxz} = 3\cos\theta\, \sin\theta\, \cos(\phi)$

$Y_{dyz} = 3\cos\theta\, \sin\theta\, \sin(\phi)$

$Y_{dx2\text{-}y2} = 3\sin^2\theta\, \cos(2\phi)$

$Y_{dxy} = 3\sin^2\theta\, \sin(2\phi)$

16.4 ONE ELECTRON IONS

Hydrogen is not the only "atom" with one electron. Atomic ions such as He^+, Li^{++}, and Be^{+++}, also have only one electron, and so their electron is also represented by the functions in table 16.1, with one important difference: since the nucleus has greater charge (Z), then the electron is drawn closer to the nucleus. This change can be introduced by replacing each occurrence of (r/a_o) in the radial functions R_{nl} with (Zr/a_o). This change will simultaneously cause the radial nodes to move closer to the nucleus and the exponential function to decay more rapidly with r. Thus, the one-electron ion orbitals look exactly like the hydrogen atom orbitals, but are compressed toward the nucleus commensurate with the nuclear charge Z.

16.5 FULL 3D SOLUTIONS $\psi(r,\theta,\phi)$

The full solutions to the 3D TISE are simply R(r) times $Y(\theta,\phi)$ as seen in table 16.1. These solutions are the orbitals (1s, 2s, $2p_x$, $2p_y$, 3s, $3p_x$, $3p_y$, $3p_z$, $3d_z2$, etc). The first few orbitals of lowest energy are plotted in table 16.2, where the drawn contours enclose all points for which the probability is higher than a certain value. Bear in mind that the electron can be found at points outside of the contours, but it is less likely. The orbitals should therefore be viewed not as hard surfaces, but as fuzzy distributions that fade to zero. We also recognize the presence of radial nodes as discussed in the previous section and in problem 16.1.

For instance, only 2p orbitals are simple (fuzzy) dumbbell shapes. The 3p orbitals are segmented dumbbells, while 4p orbitals have more segments, like some sort of insect. These radial nodes are in addition to the angular nodes discussed in Chapter 15. In fact, the total number of nodes, radial plus angular, depends only on the value of the principal quantum number ***n*** (see problem 16.2).

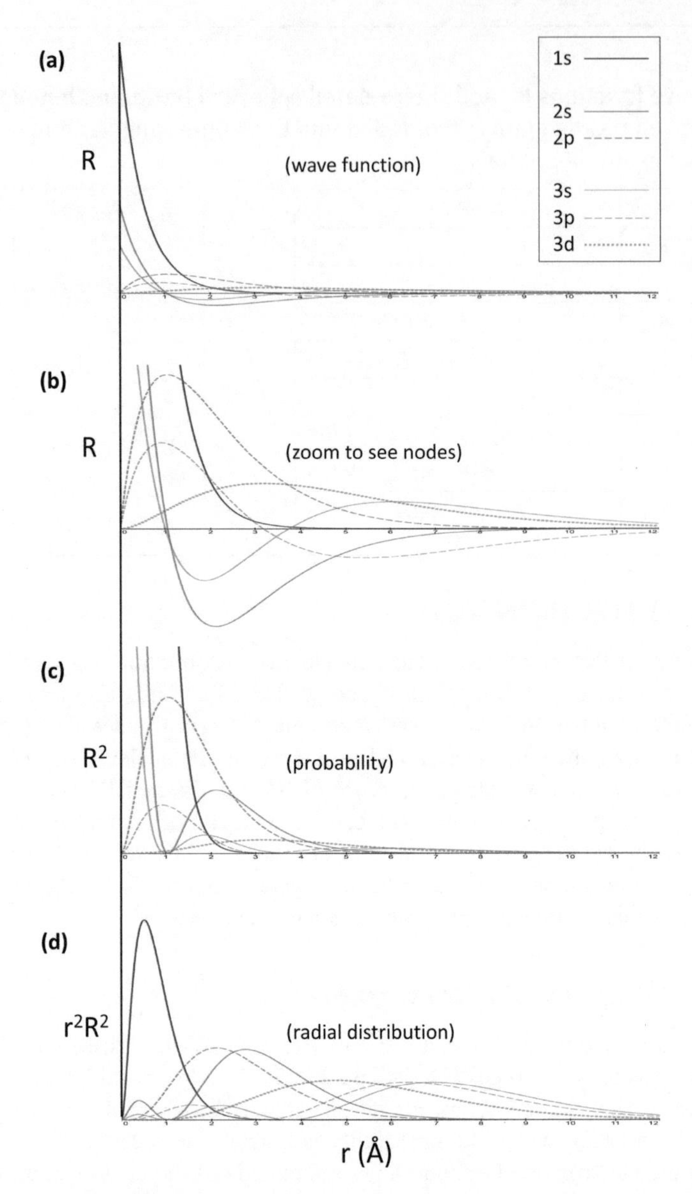

FIGURE 16.1 Normalized hydrogen atom radial wave function R_{nl} for n = 1, 2, 3. **(a)** Radial wave functions R_{nl}; **(b)** zoom in on y-axis to show node positions; **(c)** radial probability, found by squaring R_{nl}; **(d)** radial distribution function, found by multiplying the probability by r^2. See key in upper right-hand corner.

TABLE 16.2

Orbitals and quantum numbers. The quantum numbers *n*, *ℓ*, and *m* are tabulated for orbitals in all seven rows of the periodic table. For instance, the quantum numbers $n = 3$, $\ell = 2$, $m = \pm 2$ correspond to a $3d_{x^2-y^2}$ orbital. The table entries "..." merely mean "etc.": the orbitals for these entries are not pictured. [Image by © Dhatfield, https://commons.wikimedia.org/wiki/Category:Atomic_orbitals_(3D_blue_red), licensed under CC BY-SA 3.0.]

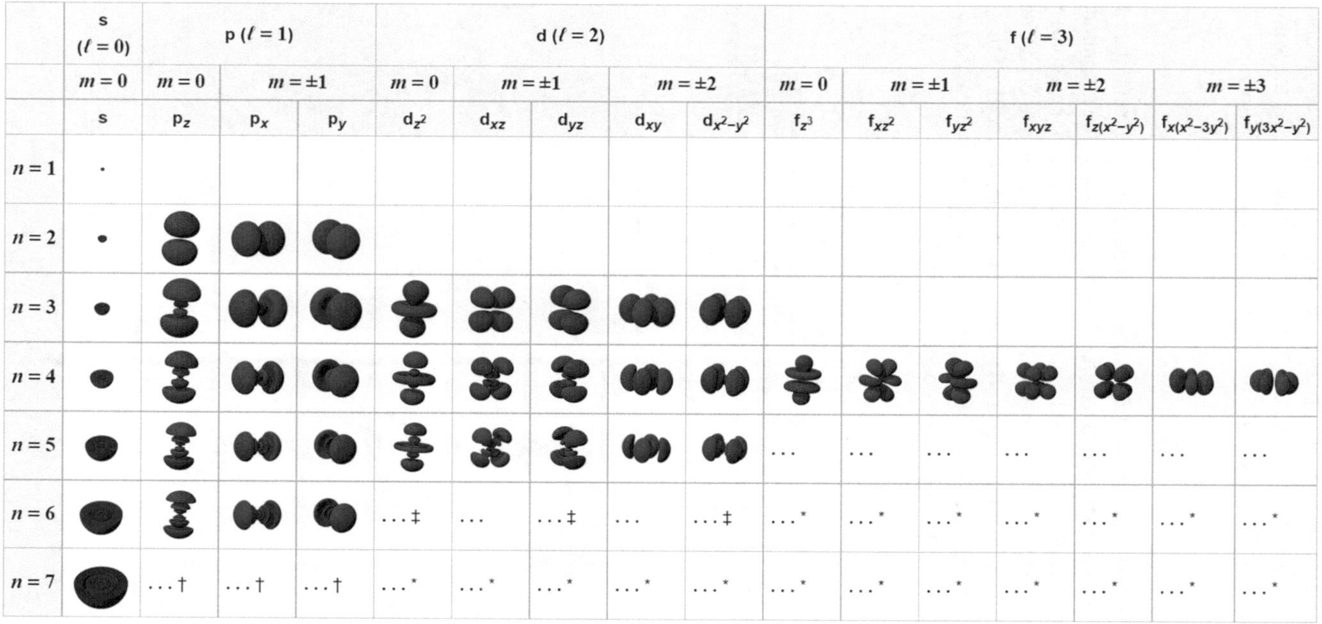

	s ($\ell = 0$)	p ($\ell = 1$)			d ($\ell = 2$)					f ($\ell = 3$)						
	$m = 0$	$m = 0$	$m = \pm 1$		$m = 0$	$m = \pm 1$		$m = \pm 2$		$m = 0$	$m = \pm 1$		$m = \pm 2$		$m = \pm 3$	
	s	p_z	p_x	p_y	d_{z^2}	d_{xz}	d_{yz}	d_{xy}	$d_{x^2-y^2}$	f_{z^3}	f_{xz^2}	f_{yz^2}	f_{xyz}	$f_{z(x^2-y^2)}$	$f_{x(x^2-3y^2)}$	$f_{y(3x^2-y^2)}$
$n = 1$																
$n = 2$																
$n = 3$																
$n = 4$																
$n = 5$										...	...	...	...	...	...	...
$n = 6$					...‡	...	...‡	...	...‡	...*	...*	...*	...*	...*	...*	...*
$n = 7$		...†	...†	...†	...*	...*	...*	...*	...*	...*	...*	...*	...*	...*	...*	...*

16.6 ENERGIES

The energies of each orbital can be calculated in a number of ways. The most straightforward way is to use the 3D TISE itself, plugging in the solution of interest (for instance, the 1s orbital solution). This produces the following energies:

$$E_n = -13.6 \text{ eV } (Z/n)^2 \tag{16.4}$$

Hydrogen atom energies are graphed in figure 16.2. The energies are negative because they describe a trapped electron. The energy value is the amount of energy required to liberate the electron from the atom. This amount is highest for ***n*** = 1. There is no ***ℓ*** or ***m*** dependence to the energy: the energy of a 3s, 3p, and 3d orbital are all equal (this will change for multi-electron atoms). Equation 16.4 tells us that for an He^+ ion with a nuclear charge of two, the magnitude of all energies increase by a factor of four: the electron is four times more deeply trapped.

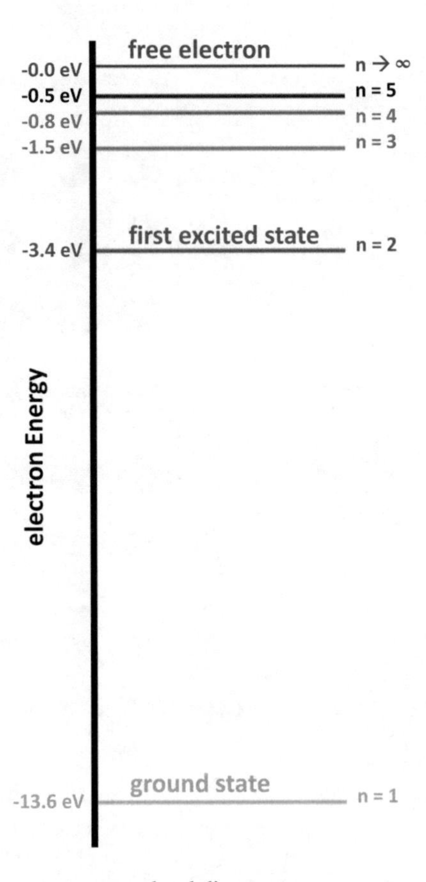

FIGURE 16.2 Hydrogen atom energy level diagram.

PROBLEMS

16.1 By inspecting figure 16.1, surmise a relationship between the shell number and the number of radial nodes.

16.2 **(a)** By analyzing a few examples, show that the total number of nodes in a full solution $\Psi(r,\theta,\phi)$ depends only on the quantum number $\boldsymbol{n}$, and not on $\boldsymbol{\ell}$ or $\boldsymbol{m}$. **(b)** Develop a simple equation relating the total number of nodes to the quantum number $\boldsymbol{n}$. **(c)** What does this dependency tell you about the relative energies of the 3s, 3p, and 3d electron in a hydrogen atom? **(d)** Do you expect this $\boldsymbol{n}$ rule to hold for the *number of nodes* in multi-electron atom orbitals? Explain. **(e)** Do you expect this $\boldsymbol{n}$ rule to hold for the relative *energies* of 3s, 3p, and 3d electrons in a multi-electron atom? Explain.

16.3 **(a)** Write the complete 1s wave function. Define each part: R(r), F(ϕ), P(θ), and give the value of each of the quantum numbers $\boldsymbol{n}$, $\boldsymbol{\ell}$, and $\boldsymbol{m}$. **(b)** Do the same for a 3d orbital.

16.4 List the names (not the functional forms) of each of the radial functions R_{nl} (there are 18) needed to describe the ground state of all atoms in the periodic table.

NOTE

1 The required polynomials are the Laguerre polynomials, discovered by Laguerre in the late 19th century as solutions to a related differential equation.

17 Introduction to Multi-Electron Atoms, Molecules, and Spectroscopy

A detailed discussion of the topics of multi-electron atoms, molecules, and spectroscopy is beyond the scope of this text. A few points will be mentioned, to illustrate that we are now well-positioned to study these complex topics:

Multi-electron atoms: (one nucleus with more than one electron)

- No exact solutions can be found, since this is a three-body problem (or four-body, five-body, …).
- Solutions can, however, be approximated using the Variational Method.
- For instance, the wave function for a 1s electron in an He atom can be approximated by using a linear combination of 1s, 2s, 3s orbitals, each with a different *effective* value of Z, which takes into account that some of the nuclear charge is shielded by the other electron. The coefficients in the linear combination, and the effective values of Z, are each allowed to vary until a minimum energy is found.
- This provides an estimate of the electron's energy and an approximate atomic orbital shape and size.
- Small perturbations, such as interactions between the angular momenta or spins of the particles, can be accounted for using TIPT to correct the energies and predict splitting of energy levels.
- In addition, the electrons repel each other, so they tend to avoid each other. This lowers the energy of each electron. This is called *interaction energy*, and it is a stabilizing effect.
- The above analysis leads to a complex energy level diagram even for simple atoms.
- The principles that we have already covered lay the groundwork for these calculations.

DOI: 10.1201/9781003356172-21

<u>Molecules:</u> (multiple nuclei with many electrons; some electrons shared)

- Assume that inner shell electrons remain in their atomic orbitals: not shared.
- In contrast, outer shell (valence) electrons mingle with valence electrons from other atoms in the molecule. Therefore, the valence orbitals mix to form "molecular orbitals".
- The mixed molecular orbitals can be approximated by taking a linear combination of the valence shell atomic orbitals from the bonding atoms. The Variational Method is used to calculate the coefficients of mixing.
- This is LCAO-MO: linear combination of atomic orbitals to create molecular orbitals.
- LCAO-MO produces approximate electronic energies and molecular orbital shapes and sizes.
- Adding versus subtracting AOs produces bonding versus anti-bonding MOs (also non-bonding).
- Bond order is found by comparing the number of electrons in bonding versus anti-bonding MOs.
- New types of motion become possible in molecules, including vibrations and rotations.
- Vibrations: bonds vibrate. That is, the nuclei act as though they are held together by springs. Therefore, SHO analysis (Chapter 10) can be used to understand bond vibrations. This includes the fact that bonds can never stop vibrating, even as $T \rightarrow 0$ K (see equation 10.9).
- Rotations: methyl groups, aromatic rings, and other groups can rotate relative to the molecule, and the entire molecule can rotate in space. Although technically an isolated atom can rotate, since the atom is spherically symmetric, the rotation has no consequence, and needn't be considered. Molecules are not spherically symmetric, so their rotations can be detected.

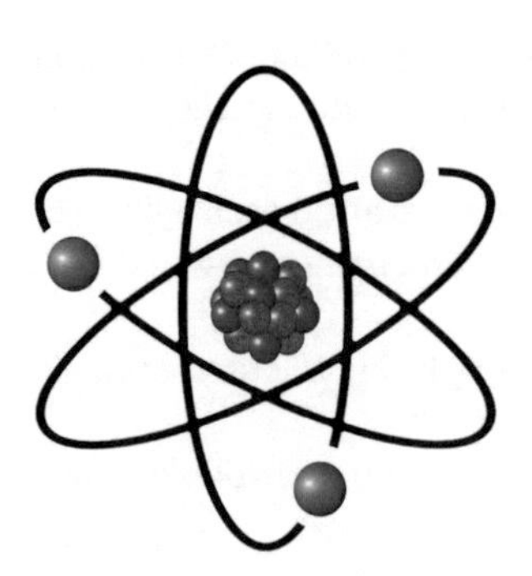

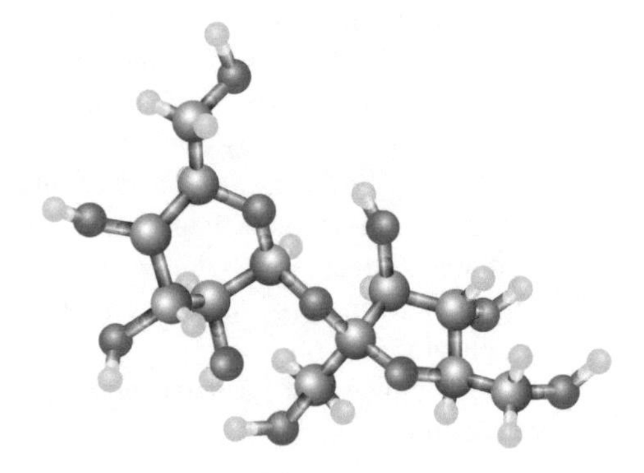

FIGURE 17.1 A multi-electron atom and a simple molecule. [LEFT image by © Gerd Altmann, https://pixabay.com/illustrations/atom-symbol-character-abstract-68866/, RIGHT image by © William Crochot, https://commons.wikimedia.org/wiki/File:Sucrose_molecule.svg, licensed under CC BY-SA 4.0.]

Spectroscopy

(Spectroscopy is used to investigate atoms or molecules by causing transitions between their energy levels. The differences in the energy levels are detected.)

Types of spectroscopy:

- UV-vis or electronic: an electron changes orbitals, e.g. from a 1s to a 2p orbital.
- IR or vibrational: amplitude of the inter-nuclear bond vibrations change (SHO).
- Microwave/rotational: rate at which the molecule or group rotates changes.
- NMR: nuclear spin state changes (similar to the spin of the electron).

From the resulting spectra, various types of information can be extracted including atomic composition, types of bonds, strength and length of bonds, and the shape of the molecule. This information is extracted from the spectra, most notably, by detecting ω_{ij}. (see equation 12.10 and the ensuing discussion), which is the frequency of light that causes the transition (absorption spectroscopy) or is given off after the transition (emission spectroscopy).

Each of these types of spectroscopy is worthy of an entire book or an entire research career. But they each trace back to the principles that we have discussed in this short treatise: each can be understood and studied using the Schrödinger equation in three dimensions, with the addition of spin. With this understanding, spectroscopy becomes not just a black box used to detect the presence of a molecule, but an extremely powerful approach to understanding the inner workings of atoms and molecules…

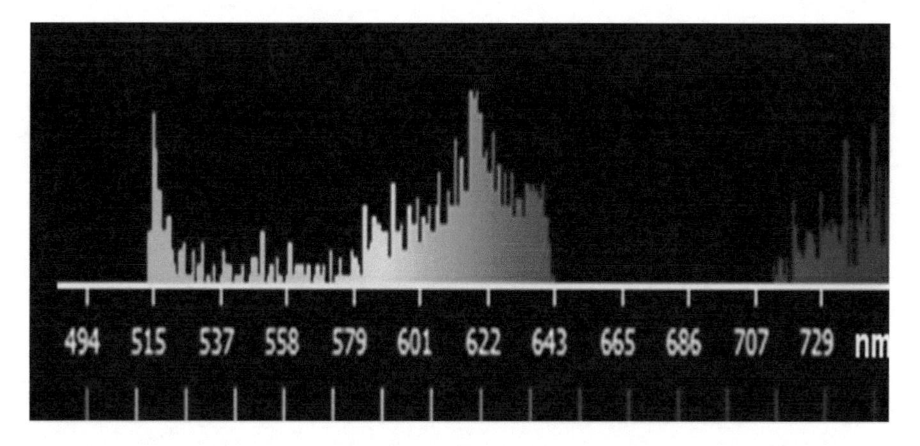

FIGURE 17.2 A colorized UV-vis spectrum. [Image by © Yurij Kot, https://commons.wikimedia.org/wiki/File:Single_cell_spectroscopy.jpg, licensed under CC BY-SA 4.0.]

PROBLEMS

Answer the following by referring to a physical chemistry textbook or the internet:

17.1 The effective nuclear charge for helium electrons in their ground state is $Z_{eff} = 1.70$. Explain what this means and name the approximation method by which the value of Z_{eff} is derived.

17.2 Briefly define the Zeeman effect. Define the perturbing Hamiltonian $\mathcal{H}_1$ that can be used to analyze this effect.

17.3 Assuming that only valence (outer shell) electrons participate in bonding, how many electrons are available for bonding in a carbon atom?

17.4 Discuss why there is no barrier to rotation of an entire molecule in a dilute gas. Given this fact, what can be said about the global angular momentum associated with the dilute gas molecule?

17.5 Which type of spectroscopy is most useful to directly find bond strength? Briefly explain.

17.6 Which type of spectroscopy is most useful to directly find bond length? Briefly explain.

17.7 What one word describes the situation in spectroscopy when the frequency of the incident light (times Planck's constant) matches a difference in energy levels in a molecule?

17.8 What do the letters "NMR" stand for? Refer to your answer to question 17.7 to surmise what each peak in an NMR spectrum represents.

Index